Behind the Scenes of the Universe

Behind the Scenes of the Universe

From the Higgs to Dark Matter

Gianfranco Bertone

OXFORD
UNIVERSITY PRESS

Great Clarendon Street, Oxford, OX2 6DP,
United Kingdom

Oxford University Press is a department of the University of Oxford.
It furthers the University's objective of excellence in research, scholarship,
and education by publishing worldwide. Oxford is a registered trade mark of
Oxford University Press in the UK and in certain other countries

Published in the United States of America by Oxford University Press
198 Madison Avenue, New York, NY 10016, United States of America

British Library Cataloguing in Publication Data

Data available

Library of Congress Control Number: 2013938579

ISBN 978–0–19–968308–6

Printed and bound by
CPI Group (UK) Ltd, Croydon, CR0 4YY

To Nadia. And to our son Francesco,
the most sweet distraction an author can hope for.

Preface

An extraordinary discovery has recently shaken the foundations of cosmology and particle physics, sparking a scientific revolution that has profoundly modified our understanding of the universe we live in, and that is still far from over.

Pioneering astronomers in the 1920s and 1930s had already noticed suspicious anomalies in the motion of celestial bodies in distant galaxies and clusters of galaxies, but it wasn't until the late 20th century that the scientific community was confronted with an astonishing conclusion: the universe is filled with an unknown, elusive substance that is *fundamentally different* from anything we have ever seen with our telescopes or measured in our laboratories.

It is called *dark matter*, and it constitutes one of the most pressing challenges of modern science. Its existence is much more than an academic curiosity, as dark matter provides the invisible scaffolding that keeps together all astrophysical structures in the universe. Take it away from a galaxy like our own Milky Way, and all its stars and planets would fly away like bullets in intergalactic space!

In the last 30 years we have learnt a lot about the properties of this mysterious substance. For instance, we have measured its abundance in the universe with exquisite accuracy, and we now know that it is far more abundant than the matter we are familiar with. We do not yet know what its nature is, but we can confidently say that *it must be made of new, as yet undiscovered particles*, unless we are being completely misled by a wide array of astrophysical and cosmological observations.

The consequences of this discovery are astonishing. The extraordinary show offered by the cosmos—the dance of planets around stars, the delicate appearance of distant nebulae, the violent clash of giant galaxies, and ultimately the very mystery of our own existence—takes place on a colossal, invisible stage made of ever-growing halos of invisible matter. And we, as human beings, are made of a relatively rare form of matter in the universe: we are *special*, in a way we had never suspected.

A worldwide race is under way to identify dark matter, with experimental facilities that include CERN's particle accelerator and a variety of astroparticle experiments located underground and in orbit around the Earth.

In this book, aimed at general readers with an interest in science, I describe the strategies proposed by physicists to go *behind the scenes* of the universe, in order to identify the nature of dark matter. I will argue that we are about to witness a pivotal paradigm shift in physics. Thirty years have passed since the current leading theories were proposed to solve the dark matter problem, and at least two generations of physicists have worked out detailed predictions for a wide array of experimental searches and built ever-larger and more sensitive experiments to find dark matter particles.

The time has now come either to rule out these theories, and move on to new ideas, or to discover dark matter and pave the way to a new era of cosmology and particle physics.

Gianfranco Bertone

Acknowledgments

In the writing of this book, I have benefited greatly from the precious insights and useful information provided by many colleagues, including some of the pioneers who led the dark matter revolution in the 1970s, and those who laid the basis for dark matter searches in the 1980s. I wish to thank in particular Albert Bosma, David Cline, John Ellis, Adam Falkowski, Jonathan Feng, Carlos Frenk, Dan Hooper, Rocky Kolb, Jerry Ostriker, Jim Peebles, Bernard Sadoulet, Joe Silk, Alar Toomre, Juri Toomre, Scott Tremaine, and Michael Turner for very interesting discussions that ranged from simple fact-checks to long discussions on the history of physics and astronomy.

Special thanks go to Roberto Trotta, for many useful comments on the manuscript; to Auke-Pieter Colijn, for suggesting the parallel between the Michelson–Morley experiment and the "nightmare scenario" discussed in the text; to Mark Jackson and Miguel Pato for a careful reading of the manuscript and useful comments; and to Lorenzo Tripodi, for many inspiring conversations. Thanks also to Oxford University Press and, in particular, to Sonke Adlung for believing in this project, and for the support from the initial proposal to publishing.

I am indebted to the many friends, colleagues, postdocs, and students I had the pleasure to work with, including Donnino Anderhalden, Chiara Arina, Ted Baltz, Lars Bergstrom, Enzo Branchini, Torsten Bringmann, David Cerdeno, Marco Cirelli, Jan Conrad, Juerg Diemand, Joakim Edsjo, Mattia Fornasa, Fabio Iocco, David Merritt, Georges Meynet, Aldo Morselli, Miguel Pato, Lidia Pieri, Roberto Ruiz de Austri, Guenter Sigl, Joe Silk, Hamish Silverwood, Charlotte Strege, Tim Tait, Marco Taoso, Roberto Trotta, Christoph Weniger and Andrew Zentner; these people took part in countless conversations and brainstorming sessions that played an important role in shaping the ideas discussed in this book.

I acknowledge the support of the institutes and funding agencies that made the collaboration with these extraordinary scientists possible. I am happy in particular to acknowledge the support of the European Research Council through a Starting Independent Research Grant and of the GRAPPA Institute of the University of Amsterdam. I would also like to thank the director and the staff of the Kavli Institute of Theoretical Physics for their hospitality and the financial support made possible by their Simons Foundation Grant during the final stages of this project.

A warm thank-you to my family, especially to Antonio, Nadine, Sophie, Alex and my parents, for consciously or unconsciously providing many suggestions for this book.

Finally, I thank my wife, Nadia, without whose constant support, encouragement, patience, and invaluable suggestions—the most important one being perhaps that of "start writing"—this book would have never been written. And thank you to our son Francesco, who was born three months after this project was started, providing the most sweet distraction an author can ever hope for.

Text copyright acknowledgements

Contents

1

——

A dark mystery

On December 12, 2010, high-ranking members of the State Council of China met with representatives of Tsinghua University, Beijing, in the mountainous province of Sichuan, to inaugurate the world's deepest underground laboratory for astroparticle physics: a 1400-cubic-meter hall excavated deep in the heart of Jinping Mountain, two and a half kilometers below the surface.

Thirteen thousand kilometers south, almost simultaneously, a team of scientists were pumping hot water into the ice of the Antarctic plateau to drill the last of a series of 80 holes, 2500 meters deep and 60 centimeters wide, and to deploy in that hole the last string of detectors for the IceCube telescope, marking the end of its construction phase, which had started seven years earlier, and the beginning of the experiment itself.

In an even more unlikely location, on May 19, 2011, a team of astronauts aboard NASA's Space Shuttle *Endeavour* removed the six-ton detector AMS-02 from the Shuttle bay and handed it over to the astronauts of the International Space Station, in a spectacular delivery between robotic arms 350 kilometers *above* the ground.[1]

These are but the latest additions to an ever-growing list of impressive scientific facilities for astroparticle physics, a booming field of scientific research at the interface between astrophysics and particle physics: gamma-ray and antimatter satellites, orbiting

[1] Spectacular footage of the delivery can be found on YouTube at <http://www.youtube.com/watch?v=RqksBepilVs>.

freely around the Earth or attached to the International Space Station; neutrino telescopes, buried in the ice of the South Pole or anchored at the bottom of the Mediterranean Sea; and particle accelerators, smashing elementary particles at extremely high energies.

Scientists hope that by combining data from all these experiments they will be able to shed light on one of the biggest open problems of modern science, a mystery that challenges our understanding of the universe and of our place in it: *dark matter*. The roots of this mystery run deep in time, but only very recently has the dark matter problem manifested itself in all its inexorable, fierce difficulty, shaking the foundations of cosmology and particle physics.

The understanding of the universe had proceeded rather linearly from the beginning of the 20th century, when Hubble had discovered the expansion of the universe. But when, in the 1970s, scientists tried to put together the many pieces of the cosmic puzzle (like the abundance of light chemical elements in the universe, the motion of stars in galaxies, and the velocity dispersion of galaxies in clusters) and to come up with a consistent cosmological model, these pieces just didn't seem to fit. To complete the puzzle, the existence of a new form of matter, *dark* matter, had to be postulated.

In an article that appeared on October 1, 1974, in the prestigious *Astrophysical Journal*, renowned Princeton cosmologists described the paradigm shift that was taking place with these shocking words:[2]

> There are reasons, increasing in number and quality, to believe that the masses of ordinary galaxies may have been underestimated by a factor 10 or more.

A factor of ten! All of a sudden, the familiar galaxies that had been observed and studied for decades, the *ordinary* systems of stars and gas whose structure was believed to be well understood,

---

[2] J. P. Ostriker, P. J. E. Peebles, and A. Yahil, "The size and mass of galaxies, and the mass of the universe", *Astrophysical Journal* 193 (1974) L1–L4.

had become uncomfortably big, too massive, and, overall, frankly bizarre. Nobody knew anymore what they, or at that point anything else in the universe, were actually made of. The standard picture of a galaxy, thought to be a simple disk of stars rotating along with their planets around a common center, and immersed in a sea of dilute gas, had suddenly become inaccurate and misleading. Gas and stars were just a small part of a much larger, more massive, *halo* of invisible matter.

The implications of this paradigm shift are staggering, and so deep that we have just started exploring them. The most important, perhaps, is that the existence of stars, black holes, supernovae, planets, and the Earth itself, in short, everything we know, is possible thanks to a sort of "cosmic scaffolding" made up of dark matter. Take dark matter away from a galaxy, and its stars and planets would break loose like bullets in the intergalactic space. This also means that we, as human beings, are not made of the same stuff that most the universe is composed of: we are *special*, in a way we had never suspected.

In modern cosmology, dark matter provides, in a way, the "stage" for the "cosmic show", a stage that was assembled when the universe was young, way before the time when stars started to shine and planets started to form, and this stage is still evolving. It is, in short, *the supporting structure of the universe*. It solves in a single stroke many problems in astrophysics and cosmology, and it provides a self-consistent framework for the structure and evolution of the universe.

Physicists, however, are hard to convince. Above all, we are reluctant to introduce new concepts, let alone new forms of matter, without hard, incontrovertible evidence. As much as astrophysical observations point to the existence of this unknown component of the universe, we simply cannot accept it until we can measure its properties and study it in our laboratories. As Robert Pirsig put it in his *Zen and the Art of Motorcycle Maintenance*,

> The real purpose of scientific method is to make sure Nature hasn't misled you into thinking you know something you don't actually know.

From modifications of gravity to new particles, and from faint stars to mirror worlds, the list of solutions proposed to the dark matter puzzle is very, *very* long, and new ideas continue to be proposed today.

As a result, to the untrained eye, modern physics journals sometimes evoke those ancient manuscripts, such as medieval bestiaries or Egyptian papyri, in which real animals seamlessly mingled with bizarre monsters and other imaginary beings. In fact, they teem with a wide array of exotic particles with bizarre names and even more bizarre properties, mingling with the already rich and heterogeneous zoo of known particles and fields.

Physicists, in a sense, are like ancient geographers, who drew monsters and other imaginary beings beyond the frontiers of the known lands. In order to explain dark matter, they have imagined new particles populating the "terra incognita" of particle physics. To detect these dark matter candidates or to rule them out is one of the greatest scientific challenges of the 21st century.

Perhaps they will turn out to exist only in the imaginative minds of particle physicists, as happened for the mythological monsters of ancient civilizations. But just as some of those ancient monsters actually turned out to be distorted perceptions of real animals, the hope is that through carefully designed experiments we will finally be able to detect dark matter particles and to shed light on some of the darkest mysteries of the modern science.

This book is about the quest for dark matter: the reasons that push scientists to believe that it exists, the theories that have been put forward to explain it, and the worldwide race currently in progress to identify it.

I will argue that we are about to witness a revolution in this field of research in the next five to ten years, for either we will find dark matter particles, therefore opening up an entirely new branch of science, or we will inevitably witness the decline of the leading theories, forcing us to revisit our description of the universe.

A peek behind the scenes

If you look at the sky on a dark night, you will see the Milky Way as a glowing band of light across the firmament. The realization that what James Joyce called

> infinite lattiginous scintillating uncondensed milky way

is just the disk of stars and gas of the disk galaxy we live in, *as seen from the inside*, is for many a source of awe and inspiration, and for some a source of shivers of excitement down the spine. It gives the whole sky a sense of perspective, providing depth to the otherwise two-dimensional vault of the heavens.

Looking towards the constellation Sagittarius, you'll be looking at the Galactic center, which is at the same time the center of the disk of stars and gas of our galaxy, which constitutes essentially everything you can see in the sky with the naked eye, and the center of a spheroid of dark matter, the halo, about ten times larger, and ten times more massive than the disk.

The density of this halo is relatively high. For every square centimeter of the page you are reading, there are about 30 thousand dark matter particles passing through that surface every second, at a speed of about 100 kilometers per second.[3] The reason why we don't perceive these particles, even if we are constantly bombarded by them, is that they interact very weakly with ordinary matter, which is also the reason why dark matter is so difficult to measure.

The Milky Way contains many *substructures*, smaller concentrations of dark matter with small proportions of stars and gas, some of which are even visible to the naked eye, like the Magellanic Clouds. But the biggest concentration of dark matter

---

[3] The actual constraint derived from observation is on the mass density of dark matter, which is about 0.3 GeV cm^{-3}. To derive the number of particles through the surface, I have assumed that the mass of a dark matter particle is approximately 100 GeV, that is, about a hundred times the mass of a proton.

Figure 1.1: The Andromeda Galaxy, our sister galaxy.

beyond the Milky Way is our sister galaxy, the Andromeda Galaxy (Figure 1.1).

We can see it even with the naked eye as a blurred star, in the group of constellations that are named after the Perseus myth. But a telescope image reveals it as a beautiful spiral of stars and gas, similar in size and shape to our own Milky Way. It is the most distant object we can see with the naked eye, and the only extragalactic object that can be observed from the northern hemisphere.

With a telescope image, we can do more than admire its beauty. Since Newton, in fact, we have known how to calculate the velocity of an object gravitationally bound to a given mass: the faster the celestial object, the larger the mass. We can calculate, for instance, the speed of the Earth and of all the other planets as they orbit around the Sun, given the size of their orbits. But if we apply this technique to calculate the velocity of the stars in the Andromeda Galaxy, it fails.

This had actually already been observed by Horace W. Babcock in 1939: the Andromeda Galaxy rotates very fast at large radii, *as if* most of its mass lies in its outer regions. A few years earlier, a Swiss astronomer working in California, Fritz Zwicky, had observed a cluster of about 1000 galaxies in the constellation Coma Berenices with the 100-inch telescope at Mount Wilson, the same telescope that Edwin Hubble used to prove the expansion of the universe about ten years earlier.

In the paper describing his findings, after complaining about the light pollution from the city of Los Angeles (in 1931!), Zwicky noted that the galaxies in the cluster had a rather high velocity dispersion, and concluded that the Coma Cluster seemed to contain much more mass than could be inferred from visible galaxies:[4]

> If this would be confirmed we would get the surprising result that dark matter is present in much greater amount than luminous matter.

Several other pieces of evidence provided further support for the dark matter hypothesis, until in the 1970s rotation curves were extended to larger radii and to many other galaxies, proving the presence of large amounts of mass on scales much larger than the size of the galactic disks.[5]

Today, the most impressive and direct evidence of dark matter comes from *gravitational lensing*, a well-established effect that relies on Einstein's general relativity, and is derived directly from the very idea at the heart of this theory.[6] The famous astrophysicist John Wheeler summarized the theoretical framework behind the effect with the words[7]

---

[4] F. Zwicky, "Die Rotverschiebung von extragalaktischen Nebeln", *Helvetica Physica Acta* 6 (1933) 110–127; English translation in *General Relativity and Gravitation* 41 (2009) 207–224.

[5] See the next chapter for further details.

[6] This effect was first tested for in 1919 by Sir Arthur Eddington, in an expedition to the island of Príncipe, off the coast of Africa, to watch a solar eclipse, providing a spectacular confirmation of Einstein's theory; see page 22, under "Dark scaffolding" in Chapter 2, for more details.

[7] J. A. Wheeler, *Geons, Black Holes, and Quantum Foam*, W. W. Norton (2000).

Mass tells space–time how to curve, and space–time tells mass how to move.

Large masses in particular cause space to curve significantly, so much so that light bends around them, distorting the images of distant background objects. Looking through large concentrations of mass, therefore, is a bit like looking through a distorting glass or a fishbowl. You can see what is on the opposite side, but since the optical path of the light rays is not straight, the image you see is distorted.

This powerful technique allows us to determine the mass of big structures in the universe, like clusters of galaxies. This mass turns out to be much larger than the mass of all the stars and gas in the clusters, proving the existence of dark matter. Furthermore, in at least one case, the bulk of the mass is physically offset with respect to the visible matter, a circumstance that has dramatic consequences for theories that seek to get rid of dark matter by changing the laws of gravity.[8]

Warding off the unknown

We are therefore forced to accept that there is something else in the universe besides ordinary gas and stars. We have strong evidence that dark matter cannot be made of ordinary matter; therefore new particles *must* exist, unless we are being completely misled by a wide array of astrophysical and cosmological observations.

This is where the dark matter problem transcends the boundaries of astrophysics and cosmology, disciplines that deal with the largest structures of the universe, and becomes a bigger, deeper mystery that challenges our understanding of particle physics, the discipline that studies the fundamental constituents of matter and their interactions. The so-called *Standard Model* of particle

[8] See Chapter 2 for more details.

physics provides an incredibly accurate description of all known elementary particles and their interactions, but cosmological observations are telling us that what we have observed so far is just the tip of the iceberg.

What makes the dark matter problem so cogent today is that most particle physicists believe, for completely independent reasons, that the Standard Model is not a truly fundamental theory, but rather a simplified version of it, valid in a limited range of energies. This is in analogy to the case of gravitation, for which Newton's theory is known to be a simplified, approximate version of Einstein's general relativity.[9]

Neutralinos, axions, sterile neutrinos, mirror particles, Kaluza–Klein photons, gravitinos, sneutrinos: the new theories proposed over the last three decades to extend the Standard Model provide an endless list of dark matter candidates, each with its own properties and detection strategies. Most likely only one of them, if any, will turn out to explain the dark matter puzzle.

Above, we have compared this zoo of imaginary particles to the imaginary beasts filling the margins of ancient geographical maps, beyond the frontiers of known lands, in an attempt "to ward off the threat of the unknown by naming it".[10] But the words of Jonathan Swift, bitterly mocking ancient cartographers,

> So Geographers in Afric-maps
> With Savage-Pictures fill their Gaps;
> And o'er uninhabitable Downs
> Place Elephants for want of Towns

warn us against accepting fantasy explanations for things we ignore. We need to go beyond that frontier, detect these dark matter particles, and measure their properties with the same exquisite

---

[9] This belief is not based on discrepancies with experimental results, but on theoretical arguments, such as the so-called hierarchy problem. See the discussion in Chapter 4.

[10] Anne McClintock, *Imperial Leather: Race, Gender, and Sexuality in the Colonial Contest*, Routledge (1995).

accuracy as that which has been achieved for Standard Model
particles.

To do that, we need a strategy. We need to build experiments
that maximize our chances of discovering the particular type of
particle we are looking for, just as hunters and fishermen adapt
their tools to the specific prey they are after. Fortunately, many
dark matter candidates fall into a rather broad category for which
the same tools can be used: WIMPs—an acronym invented by
Michael Turner, which stands for "weakly interacting massive
particles".

If dark matter is made of WIMPs, we should be able to detect
it. Although all search strategies so far devised have failed to provide incontrovertible evidence for dark matter particles, a new
generation of particle astrophysics experiments is about to start,
and some such experiments have already started taking data. But
how do we detect something we know nothing about?

The quest for dark matter

In the hunt for dark matter, physicists have many weapons in their
arsenal. The most powerful are perhaps particle colliders, where
elementary particles are accelerated to very high energies, and
then smashed together in the center of big detectors. The collision
converts the initial energy of the colliding particles into a host of
different particles. This type of experiment is so well understood
that the masses and interaction strengths of almost all particles of
the Standard Model are known with very good accuracy.

Particle accelerators have become so large and powerful that
the work of many thousands of scientists is required to build and
operate them. When I visited CERN, near Geneva, in 2007 for
a conference, the organizers took us to visit the ATLAS detector just before it was completed, down in the 27-kilometer ring
of the Large Hadron Collider (LHC). I had prepared for my visit
by studying the structure and the detection principles of its various components, but I was still surprised by its sheer size: a

46-meter-long giant, as tall as an eight-story building, barely fitting into its underground hall, like the green apple in Magritte's painting *The Listening Room*.

Robert Wilson, founding director of the Fermi National Laboratory, near Chicago, famously compared the construction of accelerators to that of cathedrals. Leon Lederman, director emeritus of the same laboratory, added[11]

> Both cathedrals and accelerators are built at great expense as a matter of faith. Both provide spiritual uplift, transcendence, and, prayerfully, revelation. Of course, not all cathedrals worked.

The witty prose of Lederman captures not only the awe and excitement inspired by scientific endeavors, but also the unspoken fear haunting those who participate in challenges of this size: failure. Scientists working at the LHC have fortunately exorcised this fear, sparked in their case by an infamous accident in 2008, demonstrating the complete reliability of the experiment.

In order to identify dark matter particles at the LHC, we need first to produce them, through the collision of beams of high-energy protons, and then to search for them among the debris produced in the collisions. Producing new particles in a collision is a bit like splitting a pack of cards into many subsets. The number of cards in the pack corresponds to the total energy available in the collision, and the smaller subsets to the new particles created. At the LHC, for example, the energy of the protons circulating in the underground ring will be brought up to seven thousand times the mass of a proton.

The reason why the initial energy must be so high is that dark matter particles most probably have a large mass, and therefore a large amount of energy must be converted into mass, following Einstein's famous equation $E = mc^2$, in order to produce them. Our dark matter particle corresponds to a hand with many cards, and the initial deck must have enough cards to produce it.

---

[11] L. Lederman, *The God Particle: If the Universe is the Answer, What Is the Question?*, Mariner Books (2006).

The limit of this detection strategy is that it is possible only if one makes specific assumptions about the nature of dark matter, and different theoretical models require different search strategies. Is it possible to perform more general searches? Particle astrophysics offers two alternatives: go deep underground, or go into space.

Underground experiments aim to detect the rare interactions of a dark matter particle with a nucleus in the experimental apparatus. If we placed our experiment in a normal laboratory, the detector would be inundated by a continuous shower of cosmic rays, highly energetic particles that penetrate the atmosphere and were discovered almost exactly 100 years ago. The dark matter signal might be there, but it is buried among these uninteresting events.

It's a bit like receiving an important phone call while sitting in a restaurant. The room is loud, you don't know the caller's voice, there is background music adding to the noise, and there are people in the room with a voice very similar to that of the mysterious caller. What do you do? If the message is important, you must walk to a quieter place where you can hear the caller's voice better and avoid all that confusion.

Similarly, in order to shield our dark matter experiments from cosmic rays and single out the rare events we are looking for, we need to bring our experiments underground, below hundreds of meters of rock. Reaching these depths implies exploiting existing infrastructure, for the cost of the experiment would otherwise be far beyond the means of the astroparticle physics community. The aforementioned Jinping laboratory, for instance, has become possible in the framework of a project of titanic proportions that will eventually lead to the construction of immense hydroelectric power stations piercing Jinping Mountain to partially satisfy China's ever-growing demand for energy.

There is yet another way of searching for dark matter, at least if it is in the form of WIMPs: indirect searches. In practice, this consists of a search for ordinary particles, like neutrinos, photons (particles of light), or other particles, produced when two dark matter particles interact with each other (or, if the dark matter particle is unstable, when it decays). Space experiments like the gamma-ray telescope Fermi and the AMS-02 detector, as well as neutrino telescopes like IceCube, will soon tell us whether this technique can actually work.

Obtaining convincing evidence from astrophysical observations has proven so far to be a very difficult task. It is in fact easy to fit almost any excess in the measured energy spectrum of photons or antimatter, at any energy, in terms of dark matter particles with suitable properties. In practice, we have enough freedom to fit almost any astrophysical observation, and features in the data from many experiments of the last five to ten years have been tentatively interpreted in terms of several different dark matter candidates, an issue that will be discussed in detail in Chapter 4.

The upcoming revolution

It is impossible to predict whether dark matter particles will be identified and, if so, how this will be done. The experimental situation is controversial, the debates at international conferences are often heated, and there is a general consensus that the entire field is at a turning point.

Perhaps we are following a red herring, like the characters in a Conan Doyle novel. If current and upcoming experiments fail to find evidence for dark matter, in the form of WIMPs or otherwise, an outcome often referred to as the "nightmare scenario", then most of our current search strategies will basically grind to a halt.

The best we could do in that case would be to learn from the absence of a signal in our detector, as Sherlock Holmes does in Doyle's novel *Silver Blaze*, where he famously displays his proverbial analytical skills in a dialogue with a puzzled police inspector:

> "Is there any point to which you would wish to draw my attention?"
> "To the curious incident of the dog in the night-time."
> "The dog did nothing in the night-time."
> "That was the curious incident," remarked Sherlock Holmes.

If instead we are on the right track, the detection of dark matter could be just around the corner, opening a new era in astrophysics, particle physics, and cosmology. Hundreds of scientists have taken up the challenge: an extraordinary experimental and theoretical effort is today in progress to discover the nature of dark matter particles.

But to better understand the importance of what is at stake, and the depth of this challenge, we need to go behind the scenes of the visible universe and understand what's lurking in the dark halos surrounding galaxies. How do we know dark matter is out there? How do we estimate its abundance and its distribution? What is the connection with new particle physics theories? What does dark matter imply about our role, as humans, in the universe? These questions will be the subject of the following chapters.

2

———

Lurking in the dark

What hurts you, blesses you.
Darkness is your candle.
Your boundaries are your quest.
                    Rumi (1207–1273)

An unusual optical phenomenon takes place on windless summer mornings at the southernmost tip of the Italian peninsula, where the narrow Strait of Messina separates the coast of Calabria from the island of Sicily (Figure 2.1). It's called the *Fata Morgana*, Italian for Morgan le Fay, the sorceress of the King Arthur legend, and to the eyes of the ancient inhabitants of this area it must indeed have appeared as an act of witchcraft, especially in a portion of the Mediterranean Sea that was thought to be infested by the mythological monsters Scylla and Charybdis.

The phenomenon is quite rare, as it requires very special atmospheric conditions, but it has been described by many eyewitnesses and celebrated by fine poets and philosophers.[12] The most ancient account of a Fata Morgana is probably that of Damascius, a Byzantine philosopher born in the 5th century AD, who describes, in a biography of his teacher Isodore, images of fighting warriors suddenly appearing at different locations on the Sicilian coast, especially around midday on the hottest summer days.

A vivid description of this exceptional sighting is provided by Father Igazio Angelucci, dean of the Jesuit college in Reggio

---

[12] One the most famous being perhaps the philosopher Tommaso Campanella (1568–1639). See his work *Adamo*, Canto VII, 48, 49.

Figure 2.1: 19th-century etching showing a Fata Morgana mirage at Reggio Calabria, Italy.

Calabria, in a letter to Leone Sanzio, a professor at the Roman College:[13]

> On the fifteenth of August 1643, as I stood at my window, I was surprised with a most wonderful, delectable vision. The sea that washes the Sicilian shore swelled up, and became, for ten miles in length, like a chain of dark mountains; while the waters near our Calabrian coast grew quite smooth, and in an instant appeared as one clear polished mirror, reclining against the aforesaid ridge. On this glass was depicted, in chiaro scuro, a string of several thousands of pilasters, all equal in altitude, distance, and degree of light and shade. In a moment they lost half of their height, and bent into arcades, like Roman aqueducts. A long cornice was next formed on the top, and above it rose castles innumerable, all perfectly alike. These soon split into towers, which were shortly after lost in colonnades, then windows, and at last ended in pines, cypresses, and other trees, even and similar. This is the Fata Morgana, which, for twenty years, I had thought a mere fable.

[13] Translation taken from Henry Swinburne, *Travel to the Two Sicilies*, London (1790).

Figure 2.2: An example of gravitational lensing: the light of distant galaxies is distorted by a galaxy cluster.

What Father Angelucci saw wasn't an optical illusion, nor a hallucination. It was a *mirage*: the same optical phenomenon that makes phantom oases appear in the desert and ghost ships float in the air, as in the old legend of the Flying Dutchman, just more extreme. As we shall see, a direct proof of the presence of dark matter in the universe is provided by the distortion of the images of distant galaxies due to the gravitational field of an intervening distribution of matter (Figure 2.2). But before discussing these *gravitational* mirages, we need first to understand the physical origin of Fata Morganas and other atmospheric mirages.

Seeing the invisible

The origin of these rare occurrences lies in the peculiar behavior of light, which, contrary to what we might naively think, does *not* necessarily propagate in straight lines, but can turn and bend as it travels through an inhomogeneous medium. We know intuitively that this is true in the case of a strong discontinuity in optical properties: those who use glasses to read these lines, for instance, know that their lenses have been carefully shaped to modify the direction of incoming light rays, bending them in order to achieve better focusing in the eyes.

A common manifestation of the bending of light is the "broken" appearance of objects partially immersed in water, like our legs in a swimming pool or a straw in a glass. But one can cite many other examples. For instance, the shape of the standard round brilliant cut of diamonds has been studied to optimize the "fire" and "brilliance" of the stones, two optical properties that encode the reflection and refraction properties of diamonds of a given shape; and Schlieren photography, invented in 1864 by August Toepler, is commonly used today in aeronautical engineering, although a search on the Internet reveals that there are even artistic uses of this technique.

The physical laws that regulate the propagation of light have been a subject of study for at least two millennia. One of the most prominent figures in the history of optics is Ibn al-Haytham, also known by his Latin name of Alhazen or Alhacen.

Little-known in Western culture outside academic circles, he was born in 965 in Basra, in what today is Iraq, near the border with Kuwait, and he was one of the most influential scientists of the Islamic Golden Age, which lasted from the 8th century to the siege of Baghdad in 1258.

Alhazen's contributions to science have been so important that he is today portrayed on the Iraqi 10 000 dinar banknote (Figure 2.3), ten centuries after his death. His monumental treatise on vision, *Kitāb al-Manāẓir*, built on the results obtained by Ptolemy and other Greek scientists around eight centuries earlier, and it represented a big leap forward in the understanding of light and vision.

Scientific giants like Newton, Fermat, Huygens, and many others contributed over the centuries to building a satisfactory theory of the propagation of light, until 20th-century physics placed it on

Figure 2.3: Ibn al-Haytham, also known by his Latin name of Alhazen or Alhacen, is portrayed on an Iraqi 10 000 dinar banknote, issued in 2005.

the firm foundation of quantum physics. For our purposes, a useful formulation is the one that was proposed by Pierre de Fermat, the French amateur mathematician famous for the so-called Fermat's last theorem, one of the most famous mathematical problems of all time.

Fermat argued, in a letter to the philosopher Marin Cureau de la Chambre dated 1662, that light takes the path that can be traversed in *the least time*. Since light propagates with different speeds in different media, Fermat's principle implies that light does not follow straight lines when it crosses an interface between two different media.

To visualize this effect, imagine that you are a lifeguard and that you have to rescue a drowning person. Since you can run faster than you can swim, the path that will allow you to reach the incautious swimmer from your chair in the shortest time is certainly not a straight line. You know intuitively that you have to run to the shore point closest to the person, and then to swim—that's why, incidentally, in the famous TV series *Baywatch* aired in the 1990s lifeguards used to run a lot along the shore, and only then swim—whereas the opposite strategy would be more appropriate if there were, say, many water scooters on the shore: you would run to the closest one and then ride it to the unfortunate swimmer.

It is easy to calculate the quickest path to the swimmer, and it
turns out to be precisely the path that light would take if the ratio
of the velocities in two media was the same as the ratio of your
running speed to your swimming speed or to the speed of your
water scooter. The fact that the light deviates from a straight line
is known as the *refraction* of light, and the slowing down of light
in different materials is parametrized by their *refractive index*. The
larger the ratio between the refractive indices of two materials,
the larger the angle that light will make as it crosses a boundary
between the materials.

The law that governs the refraction of light was already known
when Fermat came up with his *principle of least time*, and Fermat
was aware in particular of the influential work of René Descartes,
who had correctly derived the formula that relates the angles of in-
cidence and refraction to the ratio of the refractive indices. It was
with immense pride, but also with surprise, that Fermat was able
to derive Descartes' law, known today as the Snell–Descartes law,
starting from his principle. In a letter to Cureau de la Chambre,[14]
he wrote:

> But the reward of my work was more extraordinary, more unfore-
> seen, and happier than ever. For, after having run through all the
> equations, multiplications, antitheses and other operations of my
> method . . ., I found that my principle gave exactly and precisely the
> same proportion of refractions that M. Descartes had established.

Mirages are therefore readily explained as due to light refraction in the atmosphere: light rays from distant objects, for instance
the shores of a distant island, cross layers of atmosphere at different temperatures, and therefore with different refractive indices,
distorting the appearance of the objects.

In the rare cases where layers of hot air form above layers of
colder air—a very rare configuration indeed, since the lowest

---

[14] Letter of January 1 1662, in V. Cousin, *Oeuvres de Descartes*, Paris (1824–
1826), Vol. VI, pp. 485–507.

layers of the atmosphere are usually the hottest—light rays can bend so much that distant object can appear upside down or otherwise severely distorted, like the "castles innumerable" seen by Father Angelucci. Furthermore, the impression of a *moving* mirage arises from the fact the air is in constant movement, leading to an ever-changing pattern of distortion, which explains how those castles could "split into two towers, which were shortly after lost in colonnades, then windows" and so on.

It is now time to go back to our analogy with the propagation of light from astrophysical sources, and to discuss *gravitational mirages*. We have seen that light rays can bend as they cross layers of atmosphere at different temperatures, modifying the appearance of distance objects; what else could bend light rays as they travel through the intergalactic medium, in the vast space that separates us from astronomical sources? Those who are familiar with Einstein's theory of general relativity already know the answer: space–time itself.

Dark scaffolding

One of the greatest achievements of Albert Einstein was to establish, in his theory of general relativity, a relationship between the mass of a celestial object and the shape of the space and time—which are not independent from each other, but part of the same physical entity called *space–time*—surrounding it.

This is more than a conjecture; it is a *fact*. Einstein's theory has been tested with great accuracy, and it is now routinely implemented in satellite navigation systems and other tools used every day by billions of people. The first measurement that allowed a confirmation of Einstein's theory was reported by the *New York Times* of November 9, 1919 in triumphal terms:

> Eclipse showed gravity variations. Diversion of Light Rays Accepted as Affecting Newton's Principles. Hailed as Epochmaking. British Scientist Calls the Discovery One of the Greatest of Human Achievements.

Einstein had predicted that during an eclipse, the positions of stars appearing close to the edge of the Sun would change, because of the fact that light rays change their path around a massive object: the space–time surrounding the object is distorted by its mass, and the path followed by light—which, by virtue of Fermat's principle, is the path that can be traveled in the shortest time—is not a straight line anymore. This changes the positions of stars in the sky by a small, but measurable, angle. In an expedition to the island of Príncipe, on the occasion of the total solar eclipse of 1919, the famous astrophysicist Sir Arthur Eddington was the first to perform the measurement, confirming Einstein's prediction, and giving him immortal fame.

A beautiful, elegant formal analogy relates the bending of light due to optical refraction and the bending due to the presence of a gravitational field. Fermat's principle can in fact be generalized to the case of gravitational lensing, to obtain the "lens equation", an elegant formula that relates the observed bending of light from a distant background image to the mass distribution of the celestial body acting as a lens.

In practice, by looking at how the images of distant galaxies are distorted by a cluster of thousands or millions of galaxies, astronomers can determine the mass of the cluster and understand how mass is distributed in it. It is a bit like looking through a lens at a known object and trying to infer the properties of the lens by studying the distorted image.

The mass estimated by this method is invariably larger than the *visible* mass, which is in the form of diffuse gas between galaxies. From observations with X-ray telescopes, we can calculate the temperature and density of this "intracluster gas", and from these the mass of all the gas in the cluster. No matter what cluster we look at, the mass obtained from lensing observations turns out to be about six times larger than the mass of gas.

One might naively think that there could be a problem with the method used to infer the total mass from lensing observations or the method used to infer the mass of gas from X-ray observations, but the fact is that *no matter what method we use and what object we look at in the universe, we always end up finding more mass than we expect.*

The evidence for dark matter is based on almost a century of observations. We can identify three steps in the discovery of its existence:

1. *Dark matter exists.* It has long be suspected that the universe contains unseen forms of matter. Already in 1877 Father Angelo Secchi, one of the pioneers of astronomical spectroscopy, wrote about the discovery of "dark masses scattered in space"[15], and the existence of dark stars and dark gas has been subsequently discussed by many authors, including Arthur Ranyard, Lord Kelvin, Henri Poincare and many others.

   In 1922, at a time when modern cosmology was still in its infancy, and the universe basically reduced to the Milky Way, the Dutch astronomer Jacobus Kapteyn discovered that stars are not randomly distributed, but they are arranged in a coherent, flattened configuration. The mathematical model describing his findings allowed him to quantify the total mass, and comparing with the observed number of stars, he was able to obtain an estimate of the amount of dark matter:[16]

   > We therefore have the means of estimating the mass of dark matter in the universe. As matters stand at present it appears that this mass cannot be excessive.

   Shortly after Kapteyn's work, the famous English astronomer James H. Jeans argued that dark matter was even more abundant than luminous matter. By the time Kapteyn's pupil Jan Oort published, in 1932, what is now a classic analysis of the vertical motion of stars in the Milky Way, three estimates already existed of the *local* density of matter near the solar system—by Kapteyn, Jeans, and Linblad. Further evidence accumulated in the 1930s: we have already mentioned in the first chapter the pioneering work of Zwicky, who inferred the existence of dark matter from the motion

---

[15] Angelo Secchi, "L'Astronomia in Roma nel pontificato di Pio IX", Tipografia della Pace, Roma, 1877.

[16] J. C. Kapteyn, "First attempt at a theory of the arrangement and motion of the sidereal system", *Astrophysical Journal* 55 (1922) 302.

of galaxies in the Coma Cluster, and of Babcock, who noticed that peripheral stars in the Andromeda Galaxy rotate very fast around its center, as if most of its mass lies in the outer regions.

None of these extraordinary scientists, however, saw the presence of dark matter in the Galaxy as a challenge to their understanding of the universe. There was clearly a discrepancy between the luminous mass observed with telescopes and the mass inferred from dynamical measurements, but it was known that dim stars, below the reach of the telescopes available back then, existed and contributed to the total mass by an amount which was hard to assess. A particularly revealing sentence in Oort's work allows us to understand what these astronomers thought about the possible nature of dark matter. After performing a rough calculation of the contribution of faint stars to the total mass, Oort wrote:[17]

> We may conclude that the total mass of nebulous and meteoric matter near the Sun is . . . probably less that the total mass of visible stars, possibly much less.

This was, therefore, the meaning of *dark matter* in the first half of the century: faint stars, gas, and meteoroids.

2. *Dark Matter is ubiquitous, and very abundant.* In 1970 Kenneth Freeman, of the Australian National University, made an interesting observation: the rotation curves of several galaxies, which describe the velocity at which stars and gas orbit around the center of the galaxies hosting them, seemed to be incompatible with the observed distribution of visible matter. It is what Freeman wrote in the appendix to the paper he published in the *Astrophysical Journal* that earns him a special place in the history of dark matter:[18]

> There must be in these galaxies additional matter which is undetected. Its mass must be at least as large as the mass of the detected galaxy, and its distribution must be quite different from the exponential distribution which holds for the optical galaxy.

---

[17] J. H. Oort, "The force exerted by the stellar system in the direction of the galactic plane and some related problems", *Bulletin of the Astronomical Institutes of the Netherlands* 6 (1932) 249.

[18] K. C. Freeman, "On the disks of spiral and 50 galaxies", *Astrophysical Journal* 160 (1970) 811.

The work of radio astronomers—Rogstad and Shostak in 1972, using the Owens Valley interferometer, and Roberts and Rots in 1973, at the National Radio Astronomy Observatory—confirmed that several galaxies exhibit flat rotation curves, and provided compelling evidence for a significant amount of matter at large galactic radii, way beyond the stellar disk. And, in 1974, two groups of well-known cosmologists—the Estonians Jaan Einasto, Ants Kaasik, and Ennand Saar, and the Americans Jerry Ostriker, Jim Peebles, and Amos Yahil—argued that besides galactic rotation curves, a large number of observations seemed to point towards the existence of dark matter, including the dynamics of galaxy clusters, a measurement of the mass of the local group, and the stability of galactic disks.

It is difficult today to reconstruct—and perhaps even to define—the attitude of the scientific community towards the paradigm shift that was taking place in those years. It is probably fair to say that it took several years before the evidence for dark matter was accepted by astronomers and particle physicists. Many astronomers were skeptical about the accuracy of rotation curve data, owing to the possibly ambiguous interpretation of radio measurements, and many particle physicists were simply not familiar at all with astronomical observations.

Two important events in the late 1970s played a key role in convincing the skeptics, at least judging from the pattern of citations in the scientific literature and—most importantly—from the firsthand accounts of the pioneering men and women who led the dark matter revolution, and who kindly agreed to be interviewed for this book. First, in 1977 the prominent particle physicist Steven Weinberg—who would be awarded the Nobel Prize in Physics two years later—published a now classic paper with Benjamin Lee[19] in which he studied the constraints on *heavy leptons*—a sort of heavy neutrino—arising from the astronomical estimates of the total amount of matter in the universe. This study stimulated the interest of many particle physicists in cosmology. Second, in 1978, Albert Bosma, Vera Rubin, and collaborators obtained further evidence

---

[19] See also Chapter 7 for more details about Benjamin Lee and his collaboration with Steven Weinberg.

for the flatness of the rotation curves with optical telescopes for a large number of galaxies, convincing the remaining skeptics of the existence of dark matter.

3. *Dark matter is made of unknown particles.* In 1979, Scott Tremaine and Jim Gunn excluded the possibility that dark matter in galaxies is made of neutrinos, and they concluded that[20]

> We feel that the most likely remaining possibilities are small black holes or very low-mass stars, or perhaps some much heavier stable neutral particles.

Then, in the 1980s, cosmologists refined the estimate of the abundance of atomic matter produced in the early universe,[21] and showed that it did not match the observed amount of matter. The analysis of the cosmic microwave background radiation—for which the 1978 Nobel Prize had been awarded—further strengthened the conclusion that most of the matter in the universe was not in the form of ordinary atomic matter, facing theorists with a complex challenge and an extraordinary opportunity: that of understanding the nature of the missing mass of the universe. By the mid 1980s many prominent scientists, including the influential cosmologist Jim Peebles, one of the fathers of modern cosmology, were convinced that dark matter was fundamentally different from the familiar atoms and invoked weakly interacting massive particles as a solution to the dark matter problem. It did not take long until a connection was suggested between cosmological dark matter and the new particles arising from extensions of the Standard Model that particle physicists had been studying independently for completely different reasons.[22]

Today, almost a century after the pioneering work of Kapteyn, we have completed the inventory of the matter in the universe. Only a small fraction, about 1%, is in the form of stars. About

---

[20] S. Tremaine and J. E. Gunn, "Dynamical role of light neutral leptons in cosmology", *Physical Review Letters* 42 (1979) 407–410.

[21] See the discussion of Big Bang nucleosynthesis in the next chapter.

[22] See the next chapter.

Figure 2.4: The mass budget of the universe.

14% is in the form of gas—half of it inside galaxies and clusters, and the other half in a more diffuse and hard-to-detect form known as the warm–hot intergalactic medium—and the remaining 85% is simply missing: it is invisible to our telescopes, and it goes under the name of "dark matter".

It is like observing an iceberg (Figure 2.4): we see only the tip but, based on Archimedes' principle, we can estimate the fraction of the mass below the surface. Similarly, we cannot *see* dark matter, but we can estimate its abundance through astrophysical observations.

If it is so abundant, where is it? One of Gollum's riddles for Bilbo in *The Hobbit* comes to mind:

> It cannot be seen, cannot be felt,
> Cannot be heard, cannot be smelt.
> It lies behind stars and under hills,
> And empty holes it fills.

Dark matter is everywhere: it travels unnoticed through every object and through us, and it permeates the entire universe at all scales. This "dark scaffolding" acts as the supporting structure for all astrophysical structures, from dwarf galaxies to the largest superclusters, while stars and gas, much less abundant, sit comfortably at the bottom of an immense "gravitational well" of dark matter, like marbles at the bottom of a well in sand.

Like colossal hidden scenery, dark matter halos support the stage on which "star performers" like black holes, supernovae, and pulsars perform their act. At the centers of these gigantic architectures, generations of stars are born and die, and the primordial gas is reprocessed into heavier elements from which planetary systems, and therefore life, can form and evolve. It is dark matter that brings together cosmic structures, and it is dark matter that holds them together.

It is a humbling thought to realize that the elements that constitute our bodies, mainly oxygen, carbon, hydrogen, nitrogen, calcium, and phosphorus, exist only in very small quantities in the universe. As we shall see in Chapter 4, we are the result of a long chain of events that took place over the 13 billion years that separate us from the origin of the universe.

Here, in a nutshell, is the chain of events that led from the Big Bang to us. Dark matter and light elements like hydrogen and helium were produced in the first few minutes after the Big Bang. Dark matter halos then slowly grew from seed structures and merged into ever-larger systems, until gas fell under their gravitational pull and sunk to their centers. The gas then fragmented into clouds. The clouds collapsed into stars. And the stars produced all heavy elements, including those that constitute our body.

It is the ultimate *principle of mediocrity*: like children who learn, as they grow, that the world is much larger than the family, the city, and the country they are born in, we, as humans, have been progressively displaced from the center of the solar system and the center of the universe, and we now learn that the matter we are made of forms only a small amount of all the matter in the universe. But it is also the ultimate *principle of excellence*. We are like small gems buried in the depths of a dark mountain. We are *special*, in a way we could never have suspected.

Pocket universes

How do we validate our model of the evolution of the universe from the Big Bang to the appearance of life on the Earth? As with any other scientific model or theory, we can never *prove it right* (the best we can do is to *prove it wrong*, or "falsify" it), but we can at least test how well it can explain observational data.

This is, however, a daunting task, especially when one considers that even the three-body problem—the problem of describing the orbits of three celestial bodies under their reciprocal gravitational attraction—is extraordinarily difficult, and can only be solved in some simplified cases. How can we therefore hope to solve the problem of computing the reciprocal interactions of *all particles in the universe*?

The problem is that, in principle, one has to calculate for each particle the attraction of every other particle in the universe. Eric Holmberg, a ingenious Swedish scientist, found an original solution to the problem in 1941. He decided to simulate the interaction of two galaxies using 74 lightbulbs, together with photocells and galvanometers, using the fact that light follows the same inverse square law as the gravitational force. He then calculated the amount of light received by each cell, and manually moved the lightbulbs in the direction that received the most intense light.

Holmberg published his paper at a time when Europe, and soon most of the rest of the world, was in the throes of World War II. Sweden had remained neutral during the war, but the working conditions were difficult, as Holmberg himself admitted:[23]

> In the early days we usually had to work on a very small budget. We certainly were very poor. . . . I had to do all the work in person, including the rebuilding of the laboratory room and making all the electrical installations. . . . The scientist today is certainly a rather spoiled person, especially in the rich countries.

[23] The quote is taken from H. J. Rood, "The remarkable extragalactic research of Erik Holmberg", *Astronomical Society of the Pacific, Publications* 99 (1987) 921–951.

Holmberg published his paper in November 1941, shortly before the United States of America joined the war. In the following years, as some of the most dramatic events of human history unfolded worldwide, the work of many research institutes ground to a halt.

But science was meanwhile making progress thanks to the enormous resources made available to military research, especially at the Los Alamos National Laboratory, a top-secret research facility in the New Mexico desert, where the brightest minds of the time were asked to contribute to the construction of the atomic bomb.

The inherent difficulty of the problem, as well as the impossibility of performing routine tests of such an enormously destructive weapon, convinced the scientists involved in the "Manhattan Project" of the need to make full use of numerical techniques. The first calculations were done either by hand, through organized "hand-computing groups", composed mainly of the wives of Los Alamos scientists, or with electromechanical machines, programmed with punched cards. But by 1945 scientists were already performing calculations at a much faster speed, thanks to the advent of ENIAC (Electronic Numerical Integrator And Calculator), the first computer to be fully electronic, without moving mechanical parts.

The first application of computer calculations to gravitational systems was probably by John Pasta and Stanislaw Ulam in 1953. Their numerical experiments were performed on the Los Alamos computer, which by then had already been applied to a variety of other problems, including early attempts to decode DNA sequences and the first chess-playing program.

Twenty years later it was at another American agency, NASA, that the most powerful computers were running, thanks to the extraordinary space program that culminated with the landing of the Apollo 11 mission on the Moon.

Two young astrophysicists, the Toomre brothers, had access in the early 1970s to one of NASA's two IBM 360-95 computers, complete with high-resolution graphics workstations and auxiliary graphics-rendering machines—computing facilities far in advance

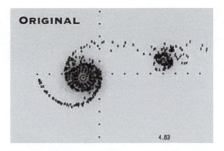

Figure 2.5: A snapshot from the movie *Galactic Bridges and Tails*: a visualization of the first numerical simulation performed by the Toomre brothers in 1972, showing the interaction of two galaxies, approximated with 120 particles.

of any other astrophysics laboratory. They set up a series of simulations of galaxy grazings and collisions using a simple code that described the galaxies as two massive points surrounded by a disk of test particles. The outcome of this analysis was a very influential paper, published in 1972, that contained a detailed discussion of the role of collisions in the formation of galaxies. Together with their paper, the Toomre brothers also created a beautiful 16 mm microfilm movie, *Galactic Bridges and Tails*[24] (Figure 2.5). When I first saw it, I found it poetically naive, and it made me think of a picture in "ASCII art" that my brother and I purchased at a local fair in the mid 1980s, with arrays of black letters and symbols forming our portrait.

Numerical simulations have improved immensely since these pioneering attempts, thanks to a dramatic increase in computing power. Modern supercomputers allow us to simulate entire universes by approximating their constituents with up to 10 billion particles, as in the case of the Millennium simulation shown in Figure 2.6, which was run at the Max Planck Society's Supercomputing Center in Garching, Germany. This has a computing power *ten million times larger* than the old IBM 360-95 used by the Toomre brothers.

---

[24] The movie *Galactic Bridges and Tails* has been recently rescanned from the 16 mm microfilm to HDTV resolution, and can be viewed at <http://kinotonik.net/mindcine/toomre/>.

Figure 2.6: A snapshot of the largest cosmological simulation ever done, the *Millennium run* (10 billion particles), which encodes the growth of cosmological structures over the entire history of the universe.

Thanks to these extraordinary tools, scientists today can create *virtual* galaxies and clusters of galaxies that very much resemble the ones we observe with our telescopes. It is by comparing the distribution of matter in numerical simulations with that observed in the universe that we can be confident of having a cosmological model that accurately describes the distribution of matter in the universe.

With an entire universe stocked on our hard drive, we can perform many interesting studies. We can, for instance, calculate how dark matter is distributed in galaxies or study the statistical properties of other astrophysical structures. It is thanks to numerical simulations that we can estimate the distribution of dark matter in a galaxy like our own, and get a handle on two crucial ingredients for dark matter searches:

- *The distribution of dark matter at the Galactic center.* As we shall see, a crucial ingredient for investigating the detectability of dark matter by astrophysical observations is its number density: the denser the dark matter distribution, the brighter it will shine. In 1996, it was discovered from a study of a numerical simulation that all dark matter halos, from those of dwarf galaxies to those of superclusters, share the same *universal* profile: the distribution of dark matter falls very steeply in the outer parts and more shallowly towards the center.[25] This specific shape of the distribution of dark matter is known as the *Navarro, Frenk, and White profile*, from the names of its discoverers. Whether dark matter precisely follows this profile is still a matter of debate and, as we shall see, this has important consequences for the study of the prospects for detecting dark matter.

- *The amount of substructures.* A quick inspection of the distribution of matter in a high-resolution numerical simulation (see, for example, the simulation of the dark matter distribution in the halo of our own Galaxy in Figure 2.7) reveals a host of small substructures, also known as *clumps* of dark matter. The fact that simulations tend to produce more substructures than are observed in the sky is considered as a potential problem for the dark matter paradigm, or at least for the *cold dark matter* paradigm, which postulates that dark matter is cold, in the sense that it possess a small velocity, and therefore it tends to cluster easily. We will see below that astrophysical processes may explain this discrepancy.

What *cannot* be inferred from numerical simulations is another crucial quantity: the *local* density of dark matter, that is, the density of particles hitting the Earth. That's because what we call a "simulated Milky Way" is nothing but a halo selected from a much bigger cosmological simulation on the basis of its resemblance, in terms of mass and size, to the Milky Way. This is useful for deriving quantities like the ones mentioned above, which are common to all halos, but certainly not for estimating with precision a crucial quantity like the local density at the Earth.

---

[25] More specifically, it follows a power law with an index of -3 in the outer parts and -1 in the innermost regions.

Figure 2.7: The Via Lactea II simulation. This is how our Galaxy would look if we could see dark matter.

A wide array of observations, including the motion of stars in the neighborhood of the solar system, the lensing of stars due to stars and planets between us and the center of the Galaxy, and mass models of the Milky Way, point to a consistent value that is about 0.3 GeV per cubic centimeter.[26] To collect one gram of dark matter, we would have to put together all the particles contained in a volume comparable to that occupied by the entire Earth!

If you think this is too little to have any influence on the motion of planets and stars, think of the vast, almost empty space between the planets or between the solar system and the next star: a quick calculation shows that the density of the planets in the solar system in a sphere that extends halfway to the next star, Proxima Centauri, is about the same as the density of dark matter!

---

[26] 1 GeV is approximately equal to the mass of a proton.

The biggest challenge in numerical simulations is implementing the effect of ordinary matter in a meaningful way. As we have seen, the distribution of dark matter can in fact be calculated simply by estimating the gravitational interactions among some particles. The size of these particles depends on the details we would like to resolve in our simulation. While a "real" dark matter particle might weigh 1000 times more than a proton, the mass of a "particle" in a numerical simulation is at least 1000 times the mass of the Sun, that is, about 57 orders of magnitude larger!

Implementing ordinary matter is hard because it does not simply sit there and interact gravitationally: it forms stars, black holes, and supernovae, it spits jets of matter into the interstellar medium, and so on. To simulate all that, we would have to resolve much smaller scales in our numerical simulations, and implement all of the physical processes underlying these phenomena. Since this impossible in practice even with the largest supercomputers, we implement these effects in an approximate fashion. In doing so, however, we lose control of the details, and the best we can do is to make educated guesses about the effects of physical processes happening on scales smaller than those resolved in the simulation.

In particular, by making suitable assumptions about the strength of the "winds" produced by the explosion of supernovae, we can explain the mismatch between the number of substructures observed in dark-matter-only simulations and the substructures observed in the Milky Way: the missing substructures are still there, but we cannot see them with our telescopes, because all ordinary matter has been blown away by supernova explosions.

An obvious question one may ask is whether the incredible growth in computing power that we have witnessed in the past four decades will allow us in the future to perform a perfect simulation of the universe, including the appearance of life.

This has long been a dream of scientists: back in the 1960s, the mathematician John Horton Conway, currently a professor at the University of Princeton, invented the "Game of Life". Although he conceived the game with pencil and paper, it is nowadays easily played on computers. The game is simple: you start with a two-dimensional grid of cells. Each cell can be either "alive" or "dead". You set up the initial conditions; that is, you decide which cells are dead and which are alive, and then the system evolves under a set of simple rules.

Despite the simplicity of the rules, the game often exhibits interesting behavior, developing self-replicating structures and moving patterns. In an excerpt from a TV program from the UK's Channel 4 that can be found on YouTube, Conway says:[27]

> My little Life game is surprising because from the simple rules one wouldn't expect to find things that move in a sort of purposeful manner. It mimics life to that tiny extent. Like a little mini-universe.

Conway's work led to intense research activity in the field of *cellular automata*, and some scientists even argued that cellular automata could lead to a new kind of science, as the title of a book by Stephen Wolfram suggests. Whether that's the case or not, it is an intriguing thought that one day we might be able to explain life with a computer simulation, which might or might not be based on cellular automata.

In fact, this has led to the extreme conjecture that we ourselves may live in a computer simulation, which is being performed by a more advanced civilization.[28] But this is a conjecture that can hardly be called scientific, since by definition we cannot perform an experiment to rule it out. Speaking about conjectures and the possibility of ruling out a theory, is it possible that the dark matter hypothesis is *wrong*?

Confucius' cat

The question is particularly cogent today, since many experiments are searching for dark matter, but we still haven't found any trace of it, and an old saying attributed to Confucius goes

[27] See <http://www.youtube.com/watch?v=FdMzngWchDk>.
[28] See N. Bostrom, "Are you living in a computer simulation?", *Philosophical Quarterly* 53(211) (2003) 243–255.

It is difficult to search for a black cat in a dark room, especially when there is no cat.

After all, the evidence for dark matter that we have discussed above relies on the strong assumption that we know the law of gravity at all scales. Is it possible that by changing the laws of gravity we could rid of this mysterious component of the universe?

In 1983, Mordehai Milgrom proposed to get rid of dark matter altogether and to replace the known laws of gravitation with the so-called MoND paradigm, short for "modified Newtonian dynamics". The price to be paid was to abandon general relativity, a theory that is particularly appealing to many physicists because of its elegance and formal beauty. But there is no dogma in physics, and history has taught us that our theories can always be refined and improved. Milgrom proposed that the law of gravity is modified below a certain *acceleration*, that is, when the gravitational force becomes very weak.

This proposal is very clever, because it bypasses one of the main difficulties of theories of modified gravity: the easiest way to construct them is to introduce a new distance scale, above which the gravitational force is modified from its characteristic inverse square law. But observations tell us that modifications of gravity (or, alternatively, the presence of dark matter) are observed on different scales in different systems.

MoND is surprisingly accurate on the scale of galaxies, and it even addresses some mysterious correlations between properties of galaxies that find no explanation in the standard dark matter paradigm. A few years later Jacob Bekenstein even embedded Milgrom's proposal into a more relativistic theory called TeVeS, for "tensor–vector–scalar" theory, promoting it from a phenomenological model to a more fundamental theory. It is without doubt an interesting proposal, and attracted and is still attracting substantial interest.

There is a problem with these theories, however: as soon as we move from the scale of galaxies to the scale of galaxy clusters, they fail to reproduce the observational data. Perhaps the

Figure 2.8: The Bullet Cluster.

biggest challenge to MoND-like theories, and one of the most direct proofs of the existence of dark matter, is provided by systems like the so-called Bullet Cluster (see Figure 2.8), a system of two clusters of galaxies that have recently collided, with one of the two passing through the bigger one like a bullet, hence the name.

By a careful analysis of the lensing of distant galaxies, it is possible to calculate the distribution of mass in the two clusters observed in the image. The mass distribution reconstructed by this technique differs from the distribution of the gas contained in the cluster. Since gas represents the main contribution to the ordinary mass of a cluster, that is, the mass made of the atoms and nuclei we are familiar with, this observation suggests that whatever constitutes the bulk of the mass in the cluster is *not* in the form of ordinary matter.

To explain this observation with MoND-like theories, one has to postulate the existence of additional matter, in the form of massive neutrinos, for instance. But there is a tension between the properties required of the neutrinos and current data, and, in general, it is not very appealing to require at the same time a modification of gravity *and* the presence of some form of dark matter.

It is still perfectly possible that it is through a modification of the laws of gravity that we will be able to explain the motion and the shape of cosmic structures, and it is important that part of the research effort of the scientific community should focus on this possibility. Fortunately, as Bekenstein says,[29]

> The increasing sophistication of the measurements in [gravitational lensing and cosmology] should eventually clearly distinguish between the various modified gravity theories, and between each other and General Relativity.

The dark matter paradigm will remain a conjecture until we finally put our hands on the particles, by measuring their properties in our laboratories. Before describing the techniques that have been devised so far to detect dark matter, however, we need to discuss what particle physics tells us about the possible nature of new particles, and how dark matter may fit into the diverse zoo of particles comprising the Standard Model of particle physics.

---

[29] J. Bekenstein, "Modified gravity as an alternative to dark matter", in *Particle Dark Matter: Observations, Models and Searches*, ed. by G. Bertone, Cambridge University Press (2010), pp. 99–117.

3

Monsters

You were not made to live as brutes but to follow virtue and knowledge
Dante Alighieri (1265–1321), *Inferno*

A familiar landscape

Choose a place that has a special meaning for you: a beautiful square in the city where you live, a seat that overlooks the ocean, a field without artificial lights on a starry night. Look around you at the variety of shapes and colors; think of the bewildering diversity of the forms of matter surrounding you.

From the small cafe where I am sitting, I can see the sculptures of Tinguely and Sainte Phalle in the Place Stravinsky, on the right bank in Paris, filling the square with movement and colors, in striking contrast to the severe facade of the medieval church of Saint Merri. I see huge white pipes, like oversized toys, delimiting the east side of the square, seamlessly sucking air into the halls of the nearby Centre Pompidou, packed with masterpieces of modern art. A street performer improvises on a catchy tune by Count Basie, while locals mix with tourists in the many cafes in the square.

Whatever the place you have chosen, there is a good chance that you have in front of your eyes a similarly representative sample of the universe we are familiar with. It seems vast, complex, and practically impossible to describe in all its details and nuances. And still, everything we see and experience, from the air we breathe to the most distant stars, from the water in the oceans to the calcium in our bones, *everything* is made of the same elementary particles: electrons and quarks.

These particles constitute the building blocks of the universe we live in. Quarks combine in triplets to form protons and neutrons (very similar in mass, but while the protons have a positive electric charge, the neutrons are neutral). The matter that we see around us is composed of "balls" of protons and neutrons glued together, which we call atomic nuclei, surrounded by a cloud of electrons. The number of protons in a nucleus identifies it as a specific chemical element.

Take carbon, for instance: its nucleus contains exactly six protons. (In fact, *by definition*, any atom characterized by this number of protons, or "atomic number", is a carbon atom. The vast majority of carbon nuclei on Earth have also six neutrons in the nucleus, but there are also isotopes with seven or eight neutrons.) To balance the positive charge of the protons in the nucleus, the same number of electrons surround the nucleus, therefore making the atom electrically neutral. The electromagnetic interactions of the nuclei and electrons of different chemical elements determine the chemical bonds that lead to the formation of chemical substances.

Carbon is actually central to all forms of life, and, in terms of mass, it is the second most abundant element in our body, after oxygen. It was not produced during the Big Bang, but only much later, when stars started to burn the primordial elements into heavier and heavier nuclei.

Primo Levi, an Italian chemist and writer, wrote a beautiful short story titled *Carbon*, in which he imagines the possible history of an atom of carbon. Levi is perhaps best known as the writer of *If This Is a Man*, a devastating account of his imprisonment in the Nazi concentration camp at Auschwitz. But Levi was also a chemist, and in 1975 he wrote a book that has been deservedly named the "best science book ever" by the British Royal Institution. Each short story is named after a chemical element, around which the narrative is centered in one way or another.

Levi starts his story from a carbon atom bound to three atoms of oxygen and one of calcium, in the form of limestone. I will not reveal too many details here, in order not to spoil the pleasure of reading this poetic and fascinating tale, but to give the reader a taste of Levi's narrative, here is a particularly delightful passage:

> [The atom of carbon] is again among us, in a glass of milk. It is inserted in a very complex, long chain, yet such that almost all of its links are acceptable to the human body. It is swallowed; and since every living structure harbors a savage distrust towards every contribution of any material of living origin, the chain is meticulously broken apart and the fragments, one by one, are accepted or rejected. One, the one that concerns us, crossed the intestinal threshold and enters the bloodstream; it migrates, knocks at the door of a nerve cell, enters, and supplants the carbon which was a part of it.

Levi was mainly interested in the chemical reactions of atoms on the Earth, so he started his story from an atom of carbon in the form of limestone, deliberately ignoring its cosmic origin. Armed with our cosmological model, we can now extend the story back to a more remote past. We could, for instance, imagine the tireless workforce of dark matter pulling gas into a large cocoon of stars. Nuclear reactions in one of the stars build large quantities of helium by fusing protons together. As helium becomes more and more abundant, new nuclear reactions kick in, allowing the fusion of three helium nuclei. Our carbon atom is born.

As the star ages, its layers become mixed as in a boiling liquid, and the carbon atom thus emerges from the depths of the stellar nucleus. The star inflates, and becomes a giant. Then it inflates more, until the outer layers are blown away into interstellar space. The carbon atom travels undisturbed through the Galaxy, until it gets captured in a protoplanetary nebula, where a pale Sun is shining its first light.

There, it is swallowed into the interior of the still-forming Earth. After millions of years, it is spitted out high into the atmosphere through an immense volcano, bound to two oxygen atoms, until it is caught in a raindrop that brings it down again, this time into the ocean. Some form of plankton—perhaps an *Emiliania huxleyi*, like the one shown in Figure 3.1—swallows it and transforms it into a beautiful calcite shell. As the *Emiliania* dies, its shell sinks

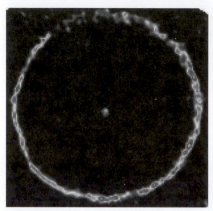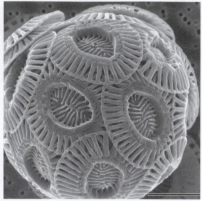

Figure 3.1: A thin shell of gas expanding from the carbon star TT Cygni, as observed with a radio telescope, and the calcite disks of an *Emiliania huxleyi* as seen through a scanning electron microscope.

to the bottom of the sea, turning with millions of other pieces of sediment into limestone, the sedimentary rock with which Levi starts his story.

Tracing a realistic story of the origin and evolution of elements in the universe has become possible only in the last few decades. But the basic building blocks were discovered much earlier. The electron, for instance, was discovered over a century ago, back in 1887, by the British physicist J. J. Thomson; the atomic nucleus was discovered in 1911 by Ernest Rutherford, the same scientist who later proved that protons were present in the nuclei of all atoms; and by the early 1930s the neutron had also been identified, whereas it wasn't until the late 1960s that the existence of quarks was proven.

Is there anything else in the universe besides simple atoms? Physicists have actually already discovered an extraordinary zoo of fundamental particles. Back in 1932, Carl David Anderson, for instance, announced his discovery of antimatter, a form of matter identical to the one we are familiar with, but with opposite electrical charge:[30]

[30] Carl D. Anderson, "The positive electron", *Physical Review* 43 (1933) 491–494.

On August 2, 1932, during the course of photographing cosmic-ray tracks produced in a vertical Wilson chamber (magnetic field of 15000 gauss) designed in the summer of 1930 by Professor R.A. Millikan and the writer, the tracks shown in Fig. 1 were obtained, which seemed to be interpretable only on the basis of the existence in this case of a particle carrying positive charge but having a mass of the same order of magnitude as that normally possessed by a free negative electron.

This discovery earned Anderson a Nobel Prize, but the new particle he discovered, called the *positron*, did not come as a surprise, since its existence had actually been predicted by one of the brightest minds of the past century, a scientist whose work has been a source of endless fascination for generations of scientists: Paul Dirac. Guided by an extraordinary physical intuition, and working from principles of elegance and beauty in the mathematical representation of reality, he discovered a relativistic formulation of Schrödinger's equation, which describes the time evolution of quantum states.

In fact, Dirac managed to bring Einstein's special relativity into quantum physics, and in doing so he realized that his equation contained a solution that didn't seem to have any physical meaning. Instead of discarding the spurious solution, he argued that this corresponded to a new, as yet undiscovered particle, and insisted that his result was[31]

> too beautiful to be false; it is more important to have beauty in one's equations than to have them fit experiment.

Many physicists share his opinion—although, admittedly, few can reach similar depths of scientific investigation—but what is important is that one can make *discoveries* in physics almost from pure thought. All theory ultimately originates from the observation of natural phenomena, but it is possible to build theories that explain these observations *and* predict new phenomena, which can in turn

[31] P. A. M. Dirac, "The evolution of the physicist's picture of nature", *Scientific American* 208(5) (1963) 47.

be tested against observations, thus completing all the steps of the scientific method.

Things started to become even more interesting in 1945, with a famous experiment that is considered by many[32] as the starting point of modern particle physics. The experiment was conducted in Rome by the three young Italian physicists Conversi, Pancini, and Piccioni, under the most unlikely conditions: Rome was occupied by the Nazis when the experiment started, in 1943, and Allied forces were bombing the city. Edoardi Amaldi, the only collaborator of Enrico Fermi who had not fled the country, provided many years later a vivid recollection of the working conditions in those days:[33]

> The target of the raid was the San Lorenzo freight yard, but more than eighty bombs had fallen on the main campus of the University, damaging several buildings. I remember that I was with Gian Carlo Wick in my office when we heard the alarm, and that, while we were running to the stairs to reach the basement, we clearly saw bombs falling on the Chemistry institute, just in front of us.

The results of this heroic effort shocked the scientific community. The history of physical discoveries seemed to be following a similar path to that of the positron: the young Japanese physicist Hideki Yukawa had predicted the existence of a "mesotron" mediating nuclear forces, and in 1936 the American scientists Neddermeyer and Anderson had found a particle that seemed to precisely match Yukawa's prediction. But the experiment of Conversi, Pancini, and Piccioni proved beyond doubt that this was *not* the case. Two years later, Yukawa's particle, dubbed the *pion*, was identified in an experiment conducted by Cecil Powell, Csar Lattes, and Giuseppe Occhialini.

The names of these scientists are hardly known outside a small academic circle. Yet, besides their fundamental contributions to

---

[32] See, for example, Luiz Alvarez's Nobel lecture.

[33] E. Amaldi, *20th Century Physics: Essays and Recollections: A Selection of Historical Writings by Edoardo Amaldi*, World Scientific (1998), p. 268.

science, they lived extraordinary lives. They lived through war and poverty. They invented new machines, started research laboratories, earned international prizes. Some of them won Nobel Prizes, and enjoyed fame and success. Others never saw their work acknowledged: they saw their collaborators winning Nobel Prizes, as in the case of Occhialini, but they were ignored, perhaps for political reasons, or perhaps because the Nobel Committee simply failed to understand the importance of their contribution.

The second half of the 20th century then became the golden age of particle physics. Thanks to an extraordinary effort that saw the collaboration of many thousands of scientists from all over the planet, a theory of all known particles and interactions was achieved, the so-called Standard Model of particle physics. This theory provides an accurate description of all particle physics experiments ever performed in our laboratories, where "accurate" means that the theoretical calculations match the measured quantities with an extraordinary precision.

Take, for instance, the so-called *fine-structure constant*, denoted by α, which characterizes the strength of electromagnetic interactions. There are many ways to measure it, but the most precise involves the use of a Penning trap, a small device that confines electric charges thanks to a combination of electric and magnetic fields. A group of researchers at Harvard has recently suspended a single electron in such a trap, and determined the "magnetic moment" of the electron. Even if you are not familiar with this quantity, the reason why it is relevant to this discussion is that in the Standard Model of particle physics you can calculate from this measurement the value of the fine-structure constant, which for historical reasons is usually given in terms of its inverse,

$$\alpha^{-1} = 137.035\ 999\ 710 \pm 0.000\ 000\ 096. \qquad (3.1)$$

Not only does this value match perfectly the previous—and completely independent—estimates, but it also provides a measurement of α with a precision of one part in one *billion*!

The Pillars of Hercules

We can broadly arrange the particles of the Standard Model into two categories:

- *Fermions* constitute all forms of the matter we know. They include the *quarks*, which, as we have seen, constitute the *nuclei* of atoms, and the *leptons*, which include the electrons surrounding the atomic nucleus; their "heavy cousins", the muon and the tau; and neutrinos.

- *Bosons* are responsible for all of the interactions between fermions: they are the carriers of forces. The quarks inside the nucleus, for instance, interact with each other by exchanging gluons, the gauge bosons that carry the strong nuclear force. Besides gluons, the other gauge bosons are the photon, which carries the electromagnetic force, and two other particles called the W and the Z, for the weak nuclear interaction. The most famous particle in this category is perhaps the Higgs boson, and we will discuss the importance of its recent discovery in chapter 5.

Simplifying a bit, we could say that *matter is made of fermions and it is glued together by bosons*. We will come back to this point when we discuss the mass budget of the universe.

Both categories include particles with a broad range of masses and a diverse array of other properties. But they differ in a fundamental quantity called *spin*, which makes them behave in fundamentally different fashions. The spin of bosons is an integer (e.g., 1 for the photon, 2 for the graviton, etc.), while fermions' spin is a half-integer ($\frac{1}{2}$ for quarks and leptons).

This apparently innocuous difference leads to macroscopically different behavior. Fermions obey the *Pauli exclusion principle*, which posits that two of them cannot share the same quantum state. The electrons surrounding the nucleus of an atom, for instance, are arranged in a hierarchy of energy levels. Rather than as a particle moving in a circular orbit, like a planet around a star, an electron is better approximated as a *cloud* around the nucleus: a spherically symmetric one, in fact, for the first energy level. The denser the cloud, the more probable it is that one will find the electron there if one performs a measurement.

The other electrons must live somewhere else, because of Pauli's principle, and they therefore become arranged in a hierarchy of states progressively less bound to the nucleus. The outermost electrons determine the chemical properties of atoms, since two or more atoms can share the "cloud" of their external electrons to create chemical substances.

The simple rule that two fermions cannot share the same quantum state therefore leads to an extraordinarily rich phenomenology. Chemistry, for instance, arises because electrons, being fermions, progressively occupy all the available energy levels surrounding the nucleus (Figure 3.2). The electron configurations around the nucleus are responsible for all chemical reactions: two ancient, per se boring hydrogen atoms can share pairs of electrons with an oxygen atom in what is called a *covalent bond* to form a water molecule. And the *hydrogen bond* between base pairs is responsible for the double-helix structure of DNA.

Bosons, in contrast, can share the same quantum state. If fermions were bosons, all electrons would collapse into the lowest

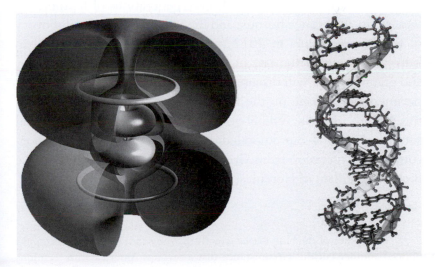

Figure 3.2: The far-reaching consequences of Pauli's principle. *Left*: visualization of electron configurations around an atomic nucleus. *Right*: The double helix of DNA, shaped by hydrogen bonds.

energy state, and chemistry, and therefore life as we know it, would not exist.

To visualize the difference between bosons and fermions, think of the behavior of car drivers on highways in different nations. In the Netherlands, for instance, cars tend to occupy as much as possible the lanes on the right, which correspond to a lower speed, or to lower levels of energy, to continue our analogy with electrons around a nucleus. In fact, in the absence of heavy traffic, cars in the furthest lane to the right drive just below the speed limit, and those in the second lane from the right slightly above the limit (but still within the tolerance limit), leaving the remaining lanes almost empty. Dutch drivers therefore behave on average like bosons, since many cars share the same speed, or the same energy level.

In the United States, the situation is quite different: cars tend to travel anywhere, especially on highways with exits on the left-hand side. They give the impression that they would gladly spread out over an arbitrary number of lanes, as if they had an urgent need to travel in their own empty, private lane: American drivers behave like fermions.[34]

This, therefore, is the Standard Model in a nutshell: matter, three generations of quarks and three generations of leptons; interactions, four gauge bosons; and the Higgs boson. The discovery of the latter particle in 2012 (yet another boson discovered in Europe!) provided the last piece of the Standard Model puzzle, and one which we will discuss in detail in the next chapter. All experiments ever done on the Earth are perfectly compatible with this model, to an incredibly good accuracy. The Standard Model sets the limits of our knowledge of fundamental physics; borrowing an expression from Greek mythology, we could call these limits the "Pillars of Hercules" of physics.

[34] The author of the blog *Résonaances* (http://resonaances.blogspot.co.uk) noticed another curious circumstance that associates America with fermions and Europe with bosons: historically, all fermions of the Standard Model were discovered in the United States, and all bosons (including, in 2012, the Higgs boson) in Europe.

Hercules carved the words *Nec Plus Ultra*, "nothing beyond this point", on the pillars he had placed on the European and African sides of the Strait of Gibraltar, which represented the limits of the known world. Dante, in his *Inferno*, celebrates the curiosity and will of Ulysses, who incites his companions to go past the Pillars of Hercules to advance knowledge and to explore what lies in the "terra incognita".

But Ulysses dies shortly after his legendary step into the unknown. Are our attempts to go beyond known physics doomed to fail, or are new discoveries awaiting us in the terra incognita of particle physics?

Terra incognita

Many theoretical physicists believe that the Standard Model is just a "low-energy realization of a more fundamental theory", or, translating from the scientific jargon, a model that provides a satisfactory description of physical phenomena only in the energy range so far explored, but is doomed to produce wrong answers when the energy of our laboratory experiments increases above a critical value.

It is a bit like general relativity. Newton's theory of gravitation provides an exquisite description of most of the gravitational phenomena we are interested in, but when applied to extreme cases, it fails. The best-known example is the anomalous behavior of the orbit of Mercury (more specifically, the precession of its perihelion), which could be explained in a satisfactory way only when Einstein introduced his theory of general relativity.

As is often the case in science, old theories are not ruled out, but simply superseded by new ones. It might seem a detail at first glance (after all, can't we simply stick to approximate theories of nature?) but it is not, since new theories always bring about a new understanding of nature. General relativity, for instance, not only is much more precise than Newton's gravity, but also provides a new *conceptual* understanding of nature, since it establishes a connection between the curvature of space–time and mass.

In the case of the Standard Model, there are many reasons to believe that it is not the ultimate theory of nature. There are three "generations" of leptons and quarks; why three? The masses of the particles don't seem to follow any specific pattern; is there a way to calculate them from first principles? Are quarks and leptons truly fundamental particles, or are they composed of even smaller constituents? Is it possible to unify all forces and interactions in one simple, elegant theory?

One of the most compelling arguments is the so-called *hierarchy problem*, which manifests itself in all its virulence in the calculation of the mass parameters of the Higgs boson. As is well known, the Higgs boson was recently discovered by use of the CMS and ATLAS detectors at the Large Hadron Collider at CERN, and its mass turns out to be around 125 GeV.

One of the pathological aspects of the Standard Model is that it predicts a much larger mass scale for the Higgs field, owing to quantum effects known as *radiative corrections*. In fact, the Standard Model calculation yields an *infinite* result, but even if one is content to have a theory valid up to some given scale, the result depends strongly on the specific value of that energy scale. This means trouble in particle physics!

Furthermore, it is the Higgs that gives mass to particles such as the mediators of the weak force, making that force, well, *weak*. We will come back to this point in later chapters. There are at least two ways out. One is to assume that the parameters of the uncomfortably large result for this mass magically conjure to make the final result small—a finely tuned and rather unnatural explanation. The other way is to work on a theory in which systematic cancellation of all the terms contributing to the Higgs mass parameter occurs.

The difference between the two approaches can be understood with an analogy. Suppose you observe from a distance someone throwing a wooden object into the air, and then you see the object circling back to the thrower. Either you can assume that the person was very lucky, and that gusts of wind have conspired to bring the object back, or you can conjecture that there is an underlying mechanism responsible for the peculiar behavior observed. For instance, the object could be a boomerang, in which case you'd have a natural explanation for the observed trajectory.

To validate your theory, however, you would have to perform an extra step by searching for specific observations, for instance the fact that the object has a V-shape, that could discriminate your theory from another (for instance, that the object is a model aircraft). This might seem trivial, but this procedure lies at the heart of the scientific method.

One *natural* method of addressing the hierarchy problem is to postulate that each particle of the Standard Model has a partner that cancels its contribution to the Higgs mass. In order for this to work, one needs to associate a boson with each fermion and a fermion with each boson, since different types of particles give contributions of opposite sign and therefore cancel each other.

This is achieved by a theory called *supersymmetry*, which emerged over the last four decades as one of the most promising theories to extend the Standard Model. The typical particle content of the theory is shown in Figure 3.3. With each quark we associate a *scalar* quark, or *squark* for short. With every lepton we associate a *slepton*, and with gauge bosons, *gauginos*.

Interestingly, supersymmetry solves many other problems of the Standard Model. For instance, it provides a framework for unifying all interactions at high energy.[35] It also predicted that the mass of the Higgs boson had to be lighter than 130 GeV—and, in fact, the discovery of this boson at 125 GeV was an enormous relief for many particle physicists, who would otherwise have been forced to trash years of theoretical effort. That does not prove supersymmetry, of course, but it is reassuring that the measurement is compatible with the prediction. What is more interesting for cosmologists, it also predicts the existence of particles with suitable properties for explaining dark matter!

---

[35] More precisely, it allows the unification of all *coupling constants*.

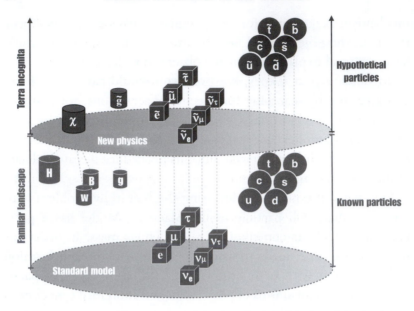

Figure 3.3: A possible extension of the Standard Model of particle physics.

The schematic setup shown in Figure 3.3 is also approximately valid for theories other than supersymmetry. If extra dimensions exist, for instance, that is, if our world is embedded in a space–time with more dimensions than what we observe in our everyday life (a possibility entertained by E. A. Abbott in his irresistible novella *Flatland*), then it is possible that the particles that we observe are just the lowest level of an infinite tower of states. In this case the partners shown in the figure would just be *excitations* of the ground state, that is, excitations of the particles of the Standard Model.

Historically, the possible existence of extra dimensions received great attention after Theodore Kaluza and Oscar Klein showed in the 1920s that adding an extra dimension to general relativity allows one to unify it with electromagnetism. Today, an additional motivation for the study of theories with extra dimensions is that string theory and M-theory, which today are the best candidates for a consistent theory of quantum gravity and a unified description of all interactions, can be formulated in theories with six or seven extra dimensions.

Going back to Figure 3.3, in this theoretical setup, the portion of the diagram where the dark matter particle might be hiding is in the upper left corner, where a particle with the symbol χ is depicted. In the language of supersymmetry, this particle is a combination of the partners of the Higgs bosons (the structure of the Higgs sector is more complicated in supersymmetry than in the Standard Model) and the partners of the electroweak gauge bosons, denoted by W and B. It is called the *neutralino*, a name that is worth keeping in mind, and which will recur often in the remainder of this book.[36]

In theories with extra dimensions,[37] the role of the dark matter particle is played by the first excitation of the B boson. To encompass this and other possibilities, the sector for the partners of the gauge bosons has been intentionally oversimplified in the figure, and it shows simply the lightest particle that can possibly appear in that sector.

The key point is that this is the sector where the *lightest* particle of the new theory (the lightest among all superpartners, or the lightest excitation of a Standard Model particle) shows up. This, and some additional properties that we will discuss in the next section, makes such a particle a perfect dark matter candidate. The trick is to make the particle stable somehow, in the sense that we do not want it to decay into anything else.

This is achieved in supersymmetry and in theories with universal extra dimensions by imposing an additional rule on the theoretical setup that states that a particle of the new theory can only decay to lighter particles if among them there is another particle of the new theory. So, for instance, a squark is allowed to decay into a quark and a neutralino, because the neutralino is lighter than the squark and is still a supersymmetric particle. But the neutralino is the lightest particle of the new theory

---

[36] There are actually four neutralinos, but it is customary to refer to the lightest neutralino as *the* neutralino.

[37] More specifically, in theories with *universal* extra dimensions.

and it therefore cannot decay into anything else: it is therefore *stable* over cosmological timescales, and whatever number of them were produced in the early universe will still be around today.

Monsters

Physicists have therefore conjectured the existence of new particles and studied the way these particles could interact with the particles of the Standard Model, but until these new particles are discovered, they will remain mere theoretical speculations.

The situation is similar to that of ancient map-makers, who guessed the shape of distant continents and their inhabitants from the accounts of pioneer travelers. In a situation where it was impossible to obtain first-hand experience, animals with an unfamiliar appearance and humans belonging to other civilizations often became imaginary beasts and monstrous human races filling the margins of ancient maps, thanks to the filter of mythological or religious interpretations or the distorting mirror of the medieval fear of the unknown.

A striking example is provided by the Psalter Map, shown in Figure 3.4. This extraordinary map was part of a medieval psalter—a collection of psalms and other prayers—and is a representation of the whole world, with Jerusalem at its center and east at the top. Asia is therefore the continent above Jerusalem, Europe is in the bottom left part of the map, and Africa is bottom right. The 14 monstrous races of humanoids shown in Figure 3.5 are drawn in the outer regions of Africa, which back then were mysterious, unexplored lands.

Echoes of the existence of these humanoid beings, described earlier by the Roman writer Pliny, and the even more fantastic accounts of bizarre creatures included in the *Travels of Sir John Mandeville*, are still present in Shakespeare's *Othello*. When the Moorish general recounts to the court his wooing of Desdemona, he alludes to his travels to Africa and Asia by referring to the

Figure 3.4: One of the most famous medieval maps, the Psalter Map, drawn in 1260 and conserved at the British Library.

Figure 3.5: Detail of the Psalter Map, showing 14 monstrous races of humanoids drawn in what then were remote, unexplored lands, roughly corresponding to the southern edge of Africa.

monstrous creatures that were thought to live in those remote lands:

> And of the Cannibals that each other eat,
> The Anthropophagi, and men whose heads
> Do grow beneath their shoulders.

To make progress, geographers had to separate reality from fiction, fairy tales from an accurate description of the world. Some of these creatures turned out to exist only in the minds of imaginative travelers in search of wonders and curiosities. But others turned out to be real, if sometimes distorted, descriptions of actual human beings or other animals.

Physics is facing a similar problem today. Our imagination is being stimulated by pioneering men and women who have found traces, in the motion of celestial objects, of a form of matter that is different from anything we are familiar with, and we are trying to figure out what it could be from a description of its properties. What "monster particles" can we expect to find once we push our accelerators to the "terra incognita" of particle physics? If we find any, can we tell what theoretical landscape they belong to?

Before tackling these questions, let us pause briefly to ask what properties a new particle should possess in order to account for all the dark matter in the universe. We will then focus on those theories, among the infinite number of possible extensions of the Standard Model, that contain "monsters" that are suitable dark matter candidates.

The following is a ten-point test that new particles must pass in order to be considered a viable dark matter candidate:

1. *Is its abundance just right?* For a new particle to be considered a good dark matter candidate, a production mechanism must exist. The problem is that for most particles the production mechanism leads to an abundance in the universe that is too large—in which case the particle must be discarded—or too little—in which case the particle can contribute only a fraction of the dark matter, as is for instance the case for Standard Model neutrinos—compared

with the abundance of dark matter measured from cosmological observations. The most promising candidates are those for which a production mechanism exists that naturally allows them to achieve the appropriate abundance in the universe, which is about 25% of all the mass–energy density of the universe, within an observational error of about 1%.

2. *Is it cold?* Dark matter particles are responsible for the clustering of all structures in the universe. If their velocity distribution is large, they cannot be confined in halos, as they tend to "free-stream" away, like water molecules from an overheated pan. If the particles are cold, that is, if they velocity dispersion is low, this does not happen; in this case, by gravitational clustering, they can grow the supporting structure for galaxies and clusters of galaxies.

3. *Is it neutral?* There are many reasons to believe that dark matter particles cannot possess an electric charge (or any other kind of charge). If they did have a charge, their interaction rate with ordinary matter would be too large. However, it is in principle possible that they might exist in the form of bound states, for instance in the form of "heavy hydrogen", where a positively charged dark matter particle is surrounded by one electron. One of the most interesting constraints arise from searches for "heavy water" in lakes and oceans; that is, a molecule characterized by HXO instead of H_2O, where one of the charged dark matter particles takes the place of one hydrogen. Null searches for these anomalous elements allow us to severely constrain this possibility.

4. *Is it consistent with Big Bang nucleosynthesis?* The set of calculations that go under the name of "Big Bang nucleosynthesis" constitute one of the most impressive successes of standard cosmology (see the next chapter). This theory predicts the abundances of the light elements produced in the first three minutes after the Big Bang; any dark matter candidate must fulfill a series of severe constraints in order not to spoil the agreement between theory and observation.

5. *Does it leave stellar evolution unchanged?* Similarly, we have achieved a rather precise understanding of stellar structure and evolution, and the agreement between theory and observation provides a powerful tool to constrain dark matter particles. Particles that could be

collected in large quantities at the centers of stars, including our Sun, could in fact lead to an accumulation of mass so large that they would collapse into a black hole, which would subsequently devour the entire star in a short time. The fact that we still observe the Sun shining allows us to rule out any combination of particle physics parameters that would destroy the Sun and other stars.

6. *Is it compatible with constraints on self-interactions?* As we have seen in the previous chapter, the Bullet Cluster provides convincing evidence that most of the mass in the two clusters in that system is dark. Interestingly, this system allows us to set a constraint on the self-interaction of dark matter particles. The "bullet" would be dragged by the larger halo in the presence of a strong self-interaction, leading to an offset between the positions of the visible galaxies (which are practically collisionless) and the total mass peak.

7. *Is it consistent with* direct *dark matter searches?* As we shall see, *direct* dark matter searches aim to detect the interactions of dark matter particles with the nuclei of a detector, usually placed deep underground. The reason for placing these experiments in underground laboratories is that they need to be shielded from the copious cosmic radiation showering from space. These experiments have made huge progress in the last four decades, and the constraints arising from these experiments are complementary to those arising from accelerator experiments.

8. *Is it compatible with gamma-ray constraints?* Alternatively, dark matter particles can be detected *indirectly* through the products of their annihilation or decay. In practice, whereas direct searches aim to detect collisions of dark matter particles with the nuclei of a detector, indirect searches aim to detect the pale light arising from the collision of two dark matter particles with each other. This process is more efficient where the density of dark matter particles is higher, and therefore we are using powerful telescopes to search for the high-energy light produced by large concentrations of dark matter lying at the Galactic center or in nearby galaxies.

9. *Is it compatible with other astrophysical bounds?* Besides gamma rays, that is, high-energy light, one can search for other particles produced by the annihilation or decay of dark matter particles, for instance neutrinos, antimatter, or light with a smaller energy than

gamma rays. There is currently a huge ongoing effort to understand the macroscopic consequences of the microscopic properties
of dark matter, and although convincing detection has not been
achieved, these techniques are useful at least for allowing us to
rule out some theoretical possibilities.

10. *Can it be probed experimentally?* Strictly speaking, this is not really
    a *necessary* condition, for dark matter particles could well be be-
    yond the reach of any current or upcoming technology. However,
    measurable evidence is an essential step of the modern scientific
    method, and a candidate that cannot be probed at least indi-
    rectly would never be accepted as the solution to the dark matter
    problem.

Physicists have proposed literally tens of possible dark matter
candidates, including neutralinos, gravitinos, sneutrinos, sterile
neutrinos, axions, fuzzy dark matter, WIMPs, WIMPzillas, su-
perWIMPs, self-interacting dark matter, cryptons, Kaluza–Klein
dark matter, D-matter, branons, Q-balls, and mirror matter, to
name a few. Behind each of these names there are bright scientists, and months, sometimes years, of dedicated work. But only
one of these candidates can actually be *the* dark matter particle:
some of them could in principle coexist in the universe, but there
is no reason why their contributions to the dark matter should
be at the same level, since these particle arise from completely
independent theoretical setups.

For our purposes, we can divide dark matter candidates into
two broad categories: WIMPs—the "weakly interacting massive
particles" introduced in Chapter 1—and non-WIMPS. The former category posits, in fact, a connection between the dark matter
problem and the "weak scale" in particle physics, while the latter
explores alternative scenarios. WIMPs can be divided into two
subcategories:

- *Natural WIMPs.* These candidates are the most widely discussed, as
  they arise *naturally* in theories that seek to address the pathologies
  of the Standard Model, in the sense that their existence was not
  predicted in order to solve the dark matter problem; instead, they

Figure 3.6: A visualization of gravitation, in a work by the Dutch artist M. C. Escher. Gravitation is visualized as being the result of the coordinated pull of monstrous creatures.

just happen to have the right properties to naturally pass the ten-point test for dark matter. Candidates in this category include the supersymmetric neutralino and the lightest Kaluza–Klein particle in theories with universal extra dimensions.

- *Ad-hoc WIMPs.* Instead of addressing the fundamental limitations of the Standard Model, one can opt for a *minimal* theory that explains "only" dark matter and nothing else. Popular candidates in this category include "minimal" and "maverick" dark matter.

For the non-WIMPs, the list is very long, and we will mention here only two of the most widely discussed candidates: axions and sterile neutrinos. Axions are hypothetical particles whose existence was postulated to solve a technical problem—the so-called strong CP problem—in the theory of quantum chromodynamics. Sterile neutrinos would interact only gravitationally with ordinary matter, apart from a small mixing with the familiar neutrinos of the Standard Model.

We will soon turn our attention to the strategies that have so far been devised to identify which—if any—of these particles constitutes the ubiquitous dark matter that pervades the universe. First, however, we need to introduce another important theme: the distribution of dark matter on astrophysical scales, and the role of this unseen form of matter in the "cosmic show".

4

The cosmic show

All the world's a stage,
And all the men and women merely players.
  W. Shakespeare (1564–1616), *As You Like It*

the powerful play goes on,
and you will contribute a verse.
  W. Whitman (1819–1892), *Leaves of Grass*

When we think of the universe, we often picture shining stars, black holes and supernovae, but most of the universe is filled with much less conspicuous forms of matter. If, inspired by the verses of Shakespeare and Whitman, we wanted to narrate the history of the universe as the plot of a "cosmic show", the "dramatis personae"—the list of characters and description of the scene that Shakespeare and many after him used to place at the beginning of their plays—would look like the one shown in Figure 4.1.

The cast would include glamorous "star performers" that attract the attention of the media and the interest of the general public: stellar explosions like supernovae and gamma-ray bursts; active black holes, like those powering quasars; and star-forming regions, like the Horsehead Nebula, which can be observed even with a modest telescope in the constellation Orion.

The role of the supporting actors, who are less known but without whom the star performers could not exist, would be played by the stuff that makes up most of the ordinary matter in the universe, like diffuse clouds of hydrogen and helium and ordinary stars. The background actors, who pass almost unnoticed in the

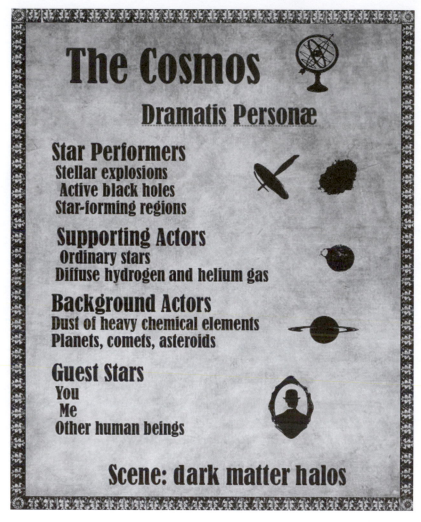

Figure 4.1: The cosmic show: scene and list of characters.

main plot although they give meaning to the screenplay, would be played by forms of matter that are irrelevant to the dynamics of the universe but central to our understanding of it, like the cold dust of heavy chemical elements, planets, comets, and asteroids.

Finally, a cameo role in this "powerful play" is played by us, human beings, who appeared on stage only at a very late time in the history of the universe. We are very similar to all other actors in the cosmic show: our blood, our brains, and our bones are made of the same form of matter that we see around us and we can see with our most powerful telescopes, just arranged in an incredibly sophisticated way.

At the same time, we like to think of ourselves more as *observers* of the cosmic show: a special form of matter fundamentally different from anything else in the universe. We know we are *alive*, in a sense which may perhaps be difficult to precisely define for philosophers and scientists, but which clearly makes us different from a rock or a drop of water. And we possess a deep form of consciousness, which makes us aware of ourselves and of the environment we live in, and which makes us different from all or at least most other forms of life.

But are we ultimately actors or observers? The theoretical physicist John A. Wheeler suggested that we should get rid of the distinction between these two roles:[38]

> one has to cross out the old word "observer" and put in its place the new word "participator".

As for dark matter, instead of including it in the cast of actors, let us assign to it the role of the stage itself, the theatrical scenes, the supporting structure that makes the cosmic show possible and allows all the actors of the "powerful play" to perform their act.

Curtain up

The universe did *not* start with an explosion. There was no "bang". The very term "Big Bang" was coined as a joke by the famous astronomer Fred Hoyle in the 1940s in a radio interview, to mock what is now the standard cosmological model theory and to promote his own "steady-state" theory, which posited that the

---

[38] J. A. Wheeler, "From relativity to mutability", in *The Physicist's Conception of Nature* (ed. J. Mehra), Kluwer Academic (1973), p. 244.

universe had no beginning and no end—a hypothesis now ruled out by observations.

Observations tell us that the universe instead underwent a phase of incredibly fast expansion, called *inflation* in cosmologists' jargon. The universe did not expand *into* something like a bomb, throwing particles and all the rest into empty space: it was the fabric of the universe itself that expanded, carrying along all its content. We do not really know what happened *before* inflation, nor even how to tackle this question meaningfully, since time itself was created with space, although many scientists have advanced hypotheses in this respect.

What we know, because we can prove it convincingly, is that immediately after inflation the universe was filled with an incredibly dense, hot soup of particles frantically colliding with each other and relentlessly destroying any emergent form of matter. To study how the universe evolved from this configuration, we refer to Einstein's celebrated equations, which, as we briefly mentioned when we discussed gravitational lensing, posit that the geometry of space–time depends on its mass and energy content. These equations are so elegant that someone, obviously under the spell of their beauty, felt compelled to paint a graffiti version of them on a rusted locomotive in a train cemetery in Bolivia (Figure 4.2).

Armed with Einstein's equations, we can describe the expansion of the universe given the specific properties of the matter and energy it contained. To study how matter looked in this early phase of the universe, scientists have been simulating the extreme conditions of pressure and temperature that existed immediately after the Big Bang at particle accelerators like the Brookhaven National Laboratory's Relativistic Heavy Ion Collider and, since 2010, at CERN's Large Hadron Collider.[39] In these colliders, nuclei of heavy elements are accelerated to very high energy and smashed together so hard that the protons and neutrons inside the nuclei "melt", releasing the quarks that constitute them and

[39] See the next chapter for a thorough discussion of the Large Hadron Collider.

Figure 4.2: Graffiti of Einstein's equations on a rusted steam locomotive in a train cemetery in Bolivia.

creating a new state of matter known as the quark–gluon plasma (Figure 4.3).

As the fabric of the universe expanded furiously, dragging along all forms of matter and energy, its temperature dropped abruptly until the first stable aggregates of fundamental particles appeared in the universe: protons and neutrons, formed when quarks and gluons are bound together in configurations of three quarks. The universe was still very young at this stage, only a millionth of a second after the start of the expansion.

Then, as the universe expanded further and cooled, the average energy of the particles decreased, making the collisions less violent, and the natural stickiness of protons and neutrons (which tend to remain glued together owing to the nuclear force) prevailed, making them aggregate together into heavier nuclei, typically made of two protons and two neutrons, to form *helium* nuclei.

The second most abundant element in the universe, helium is the gas that makes balloons stay aloft and people speak with

Figure 4.3: The primordial quark–gluon plasma is currently being studied at CERN. The two images show the tracks of particles produced in the ALICE detector from a collision of lead ions at an energy of 2.76 TeV. *Top*, a simulated event; *bottom*, an actual event recorded in 2010.

funny voices, and it got its name from the English astronomer Sir Norman Lockyer (also known as the founder and first editor of the prestigious journal *Nature*), from the Greek *Helios*, for the Sun, where its presence had been observed for the first time. Jimi Hendrix, in his last interview, days before he died in 1970, said[40]

---

[40]   Tony Brown, *Jimi Hendrix: The Final Days*, Omnibus Press (1997).

I have this little saying: when things get too heavy, just call me helium, the lightest known gas to man.

Well, helium is not the lightest gas, but it is admittedly second only to hydrogen, which is highly flammable, as the developers of hydrogen-powered vehicles know well, so it is probably a safer choice!

Elements heavier than helium could in principle have been produced, but as the universe cools and expands, the density and the probability of forming heavier elements decrease dramatically. It is the delicate balance between the rate of expansion and the rate of collisions that regulates the production of primordial elements, and the precise agreement of the calculated abundances of primordial elements with observations makes us confident that we have a reliable and precise understanding of the very first minutes after the Big Bang.

How do we know this? The details are complicated, but the main idea is simple: one has to perform a theoretical calculation and then compare the results with observations. The theoretical calculation can be set up through the implementation of a network of nuclear reactions, whose probabilities (or *cross sections*) can be derived from nuclear-physics laboratory experiments, in an expanding universe.[41]

Approximate calculations are relatively easy to perform, and yield reasonable estimates, but today these calculations are performed numerically: the only realistic way to implement a wealth of subtle effects and all known nuclear reactions. The result of the standard calculation is that 24.82% of the mass of ordinary matter is in the form of helium.

We now need to compare this theoretical result with observations. But how can we estimate the mass fraction of helium after the Big Bang from observations? The measurement is tricky because stars produce helium, and release it into interstellar

---

[41] The expansion rate itself needs to be calculated, since it depends on the contents of the universe, for instance the number of neutrino species.

space when they die, so for primordial helium we need to point our telescopes towards clouds of gas that have not been significantly contaminated by stellar formation or other chemical processes.

Once suitable regions have been identified, for instance the Orion Nebula (Figure 4.4; a beautiful object that can be seen with the naked eye south of Orion's belt), the light collected from such a region is scrutinized in search of recombination lines; these are photons with a specific energy, produced by electrons captured by helium ions that move down to a configuration (see Figure 3.2) of lower energy.

From these observations, after a few intermediate steps, it is possible to calculate what fraction of the primordial gas in these clouds is in the form of helium: the observed value is 24.9%, with an error of 0.9%, remarkably close to the theoretical estimate. Similar calculations can be performed for other primordial

Figure 4.4: The Orion Nebula as seen by the Hubble Space Telescope.

elements, and for all of them the results are consistent with the standard cosmological model.[42]

Prologue

When the universe was in its infancy, dark matter (and the other elusive component of the universe that goes under the name of "dark energy") basically did not play any role. The expansion rate was driven by *radiation*, that is, relativistic particles like photons and neutrinos. Dark matter was physically present, but it did not participate significantly in the network of reactions that led to the production of primordial elements.

This is how the influential cosmologist George Gamow, who was the first to set up a theory for the production of light elements in the early universe, described the relative abundance of photons and matter in the early universe:[43]

> One may almost quote the Biblical statement: "In the beginning there was light," and plenty of it! But, of course, this "light" was composed mostly of high-energy X-rays and gamma rays. Atoms of ordinary matter were definitely in the minority and were thrown to and fro at will by powerful streams of light quanta.

Gamow was an extraordinary scientist and, together with his student Ralph Alpher, he quickly realized that their model for the production of primordial elements (which we call "Big Bang nucleosynthesis") predicted that a relic radiation should be left over from the time when the universe was very hot.

We now know that after approximately 400 000 years a dramatic event took place: the universe had become so cool that practically all free electrons had been captured by atomic nuclei to form atoms. This was the epoch of *recombination*. The photons

---

[42] Perhaps with the exception of lithium, although the situation is still unclear.

[43] G. Gamow, *The Creation of the Universe*, Courier Dover Publications (2004).

that had been continuously scattered by electrons, just like light in a fog, could finally propagate freely, as nuclei had cleaned up all of the available electrons. Those photons have actually been propagating freely since that time, and they constitute the relic radiation known today as the cosmic microwave background.

Two Nobel Prizes in Physics have been awarded for observations of this relic emission: one in 1978 to Arno Penzias and Robert W. Wilson, who had serendipitously discovered it with a radio antenna, and another in 2006 to John Mather and George Smoot, for the discovery of *anisotropies* in the cosmic microwave background with the COBE experiment. These anisotropies are the embryos of galaxies and all other astrophysical structures in the universe.

Thanks to subsequent experiments such as Boomerang, the Wilkinson Microwave Anisotropy Probe (WMAP), and Planck (Figure 4.5), we are today able to reconstruct precisely the history of the early universe. The cosmic microwave background was released at a time when dark matter had just started growing structures, while clumps of ordinary matter "bounced" between dense and dilute configurations inside these structures. When recombination took place, photons became able to propagate freely, and they escaped from these structures with different energies,

Figure 4.5: A full-sky map of the cosmic microwave background, as measured by the Planck space telescope.

depending on how big the structures were and what path they took to reach us.

The study of the inhomogeneities in the energy of these photons has reached an extraordinary precision. The renowned Russian physicist Lev Landau famously said that "Cosmologists are often in error but never in doubt", but it is probably fair to say that the wealth of new data accumulated in the second half of the 20th century has placed cosmology on very firm ground, allowing us to enter the era of "precision cosmology".

The interested reader will find on the website of the Wilkinson Microwave Anisotropy Probe satellite a Flash applet[44] that allows anyone to adjust the parameters of the standard cosmological model to fit the data obtained by the experiment itself, which measured the power fluctuations of the cosmic microwave background. The theory behind that is rather intricate, but this simple tool allows you to determine, for instance, the amount of dark matter and dark energy required to explain the observed fluctuations in temperature.

Again, to solve the puzzle, we need to invoke the presence of dark matter, which contributes 24% of the content of the universe, while ordinary matter contributes 4% and dark energy 72%. These proportions have not been always the same, because even if all species have coexisted since the Big Bang, they have evolved differently as the Universe expanded. The behavior of the densities of radiation, matter (including dark matter), and dark energy is shown in Figure 4.6. Upon careful inspection, we see that it is only since a few billion years ago that dark energy has dominated the energy content of the universe, whereas in the past, we went through epochs where dark matter, and before that radiation, was the dominant species.

At the beginning, the largest contribution to the density of the universe was provided by relativistic particles, or radiation, as we have said. Then, the era of matter domination started,

[44] <http://map.gsfc.nasa.gov/resources/camb_tool/index.html>.

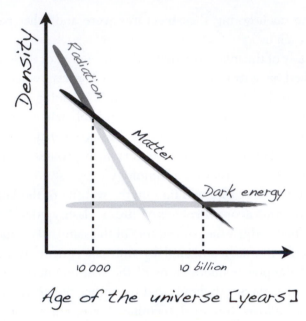

Figure 4.6: Time evolution of the different species contributing to the mass–energy budget of the universe.

which saw (dark) matter as the dominant species. Today, it is dark energy that rules over all other forms of matter and energy and drives the expansion of the universe. In fact, dark energy drives an *accelerated* expansion of the universe, as discovered by two teams of scientists led by Saul Perlmutter, Brian Schmidt, and Adam Riess, who received in 2011 the Nobel Prize in Physics for this discovery.[45]

---

[45] I hope that one of my intrepid readers will take up the challenge of going to the train cemetery in Bolivia and adding the term corresponding to dark energy to the graffiti shown in Figure 4.2. The full equation should read

$$R_{\mu\nu} - \frac{1}{2}g_{\mu\nu}R + g_{\mu\nu}\Lambda = \frac{8\pi G}{c^4}T_{\mu\nu}.$$

Note also the plus sign on the right-hand side, to match the most common sign convention.

Once matter became the dominant species in the universe, about 10 000 years after the Big Bang, big structures started to grow in the universe. It was actually dark matter that started to form these structures, since ordinary matter still consisted of a highly coupled plasma of photons, nuclei, and electrons.

One question remains to be answered, however: how was dark matter produced? One intriguing explanation becomes readily available if we assume that dark matter is made of WIMPs (weakly interacting massive particles). These particles can in fact achieve thermal and chemical equilibrium with ordinary matter at very high temperatures, through processes like the ones shown in Figure 4.7.

The density of dark matter would have decreased to a negligible amount if the reactions that turned dark matter into ordinary matter had been very efficient. But that was not the case for WIMPs – that's why they are called *weakly interacting* massive particles. When the expansion of the universe diluted the density of WIMPs so much that the reactions that turned WIMPs into

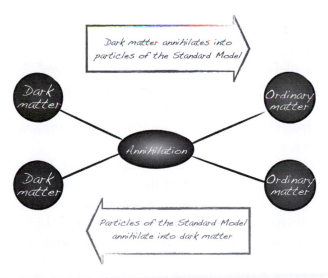

Figure 4.7: Dark matter "cold relics" are kept in equilibrium with ordinary matter by self-annihilation. The process proceeds in both directions until the rate of interactions drops below the expansion rate of the universe.

ordinary matter and vice versa could not take place anymore, the number of WIMPs remained frozen.

This mechanism, with small variations, applies to several other dark matter candidates, and we will refer generically to particles produced this way as *cold relics* of the Big Bang.

The plot

Now that we have established the setting and introduced most of the characters in the prologue, it is time to reveal the plot. We need first to clarify how all the elements that were not produced in the early universe, like carbon, oxygen, iron, and so on, form, and also how small ripples in space, as observed through the inhomogeneities in the cosmic microwave background, create immense structures in the universe, such as galaxies and clusters of galaxies.

The first act opens with 400 million years of darkness. The photons released at recombination still stream endlessly through the vast universe, but there is almost no new light being produced anywhere in the universe. The fabric of space–time is stretched, the temperature goes down, and everything seems doomed to a cold, dark silence. These are the cosmic "Dark Ages". Not an exciting start, you may think, but wait: something is lurking in the dark. The small ripples that left their mark on the cosmic microwave radiation grow larger, amplified by the tireless action of dark matter.

It's a bit like the *Black-Form Paintings* of Mark Rothko. Look, for instance, at painting No. 8 of this series, shown in Figure 4.8. It looks like solid black at first sight. But upon further inspection, you'll see texture and tone appearing, and multiple layers of color. There is substance, energy, and meaning in the painting, disguised as darkness!

Similarly, an endless though barely noticeable activity characterizes the Dark Ages, with dark matter reshaping space–time as it grows immense structures. To investigate the dynamics of this

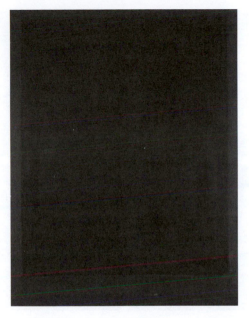

Figure 4.8: No. 8 of Rothko's *Black-Form Paintings*: a powerful metaphor for the cosmic "Dark Ages".

inaccessible portion of the history of the universe, we must resort to numerical simulations[46] (Figure 4.9).

As small concentrations of dark matter—called *halos* in cosmology—merge with each other, they shed part of their mass from the new, emerging structure. This happens because gravitational interactions induce enormous tidal waves in these systems, stripping away the external part of the systems. Larger and larger halos are built through this merger process, drawing immense quantities of gas into their deep interiors.

After 400 million years of darkness, the pressure and density of the ordinary matter, which is still in the form of primordial gas, becomes so high that they trigger a spectacular chain of events. Hydrogen atoms combine into molecules of two atoms, forming

[46] The interested reader will find more pictures and movies on the website of the Millennium simulation at <http://www.mpa-garching.mpg.de/galform/virgo/millennium/>.

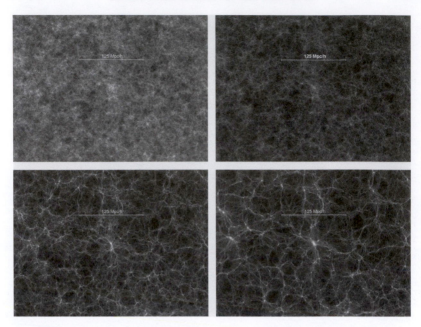

Figure 4.9: Snapshots of the Millennium numerical simulation at 0.2, 1, 4.7, and 13.6 billion years after the Big Bang, showing the formation of structures from a homogeneous universe to the appearance of halos and filaments.

molecular hydrogen, a gas that radiates energy away effectively in the form of infrared radiation, thereby cooling all the surrounding gas and allowing the formation of even denser systems.

Although current simulations cannot follow the formation of structures down to the scale of stars, it is believed that eventually the gravitational collapse of these clouds of gas is prevented by the ignition of nuclear reactions at the centers of the clouds. This marks the formation of the first stars, and the end of the Dark Ages. These objects are much bigger than our own star, the Sun, probably between 100 and 1000 times bigger, and they live very short lives.

Aging, for a star just as for us, is a matter of chemical reactions. Stars are born with a given initial chemical composition: so much hydrogen, so much helium, etc. Through nuclear reactions, however, they synthesize new elements, burning heavier and heavier elements in successive phases until they run out of nuclear fuel and dies.

Star performers

There are two possible outcomes for the first stars: they may end their life as a black hole, if their mass is between 100 and 140 times the mass of the Sun or above 260 times, or they may undergo a supernova explosion, for intermediate values of their mass. The resulting population of black holes is thought to lead, through mergers with other black holes and accretion of ordinary matter, to the formation of the *supermassive black holes* that we observe at the center of most galaxies. Our Galaxy also has its own supermassive black hole at its center. Its mass can be estimated by measuring the orbits of stars revolving around it: about 4 million times the mass of the Sun.

Figure 4.10 shows the impressive data obtained by UCLA professor Andrea Ghez and her team using the adaptive optics system of the Keck telescopes, following up on the pioneering observations of the German astrophysicist Reinhard Genzel and his collaborators. For some of these stars, a full orbit around

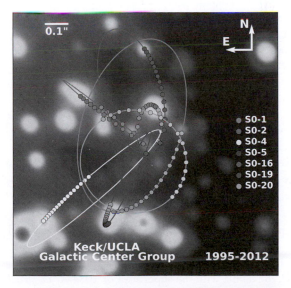

Figure 4.10: Orbits of stars around the supermassive black hole lying at the Galactic center.

the supermassive black hole has been observed, allowing a very precise determination of its mass.

One of the puzzling aspects of supermassive black holes is that some of them seem to have already been in place when the universe was relatively young.[47] This seems to suggest that they grew from already massive "seeds", rather than from stellar-mass black holes. We will return to this point in Chapter 7.

As for supernovae, these events are so violent and spectacular that even before modern telescopes could capture their beauty, their mysterious and sudden appearance fascinated countless generations of human beings. The nebula shown in Figure 4.11, known as the Crab Nebula, exploded in AD 1054,[48] and the event was recorded by several astronomers worldwide. The annals of the Chinese Sung dynasty, for example, contain an account of an observation performed by the Chief Astrologer Yang Wei-tê:[49]

On the 22nd day of the 7th moon of the 1st year of the period Chih-Ho [August 27, 1054] Yang Wei-tê said: "prostrating myself, I have observed the appearance of a guest-star, on the star there was slightly an iridescent yellow color. Respectfully, according to the disposition of the Emperors, I have prognosticated, and the result said: the guest-star does not infringe upon Aldebaran; this shows that a Plentiful One is Lord, and that the country has a Great Worthy. I request that this [prognostication] be given to the Bureau of Historiography to be preserved".

[47] Observations performed with the Sloan Digital Sky Survey have allowed us to discover the existence of quasars powered by black holes a billion times more massive than the Sun at a redshift of 7. This means that these black holes were already in place when the universe was less than a billion years old, a very short timescale for the usual theories put forward to explain the formation of these objects.

[48] More precisely, the light emitted by the explosion reached the Earth in AD 1054.

[49] Cited in J. J. C. Duyvendak, "Further data bearing on the identification of the Crab Nebula with the supernova of 1054 A. D.", *Publications of the Astronomical Society of the Pacific* 54 (1942), 91.

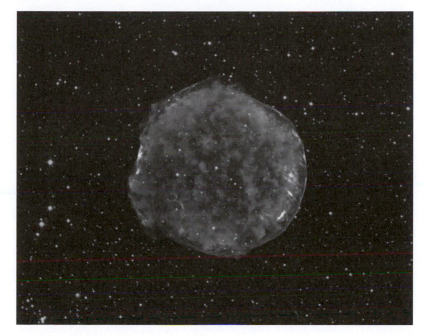

Figure 4.11: A beautiful image of the Tycho supernova remnant, as observed in X-rays with the Chandra telescope.

Back then, as today, astrologers interpreted or, more accurately, *guessed* the meaning of astronomical objects and celestial configurations, for the benefit of emperors. We know now that there was no truth in their prognostications, but we must be grateful to them for leaving an accurate record of astronomical events throughout history. Incidentally, there are reasons to be skeptical about Yang's account: the emperor's favorite color was yellow, and the real appearance of the star might have been distorted to please him.

There are many other accounts of supernova observations throughout history, and from many different civilizations, including a beautiful Anasazi petrograph in the Chaco Canyon that could actually depict the same supernova as that described by Yang Wei-tê. The estimated rate of supernovae in our Galaxy is about one every 50 years, so they are rare events on the timescale of a human life, but not on the timescale of a civilization, and even less when compared to astronomical scales.

These explosions actually play an important role in the cosmic ecosystem: they are responsible for the formation of chemical elements heavier than iron, which wouldn't otherwise be produced by nuclear reactions in the cores of stars. Every single atom of lead, uranium, and *every* element heavier than iron was produced in the violent explosion of a massive star. The entire life of a star, from the instant the first nuclear reaction happened in its core to its explosion, can be summarized as the conversion of light chemical elements into heavier ones. The fact that basically all we see when we look around us is made of heavy elements clearly means that the universe is old enough to have seen many generations of stars go by.

In an otherwise gentle and immutable cosmos, the catastrophic events that characterize the death of stars are a source of awe and wonder: the delicate appearance of the remnants of stellar explosions captured by modern telescopes and the dance of stars around the supermassive black hole at the center of our Galaxy resonate with something deep within us.

Even more striking is the connection between us and these violent phenomena, since from the ashes of these explosions, after a seemingly unlikely chain of events, life started to evolve on the Earth. And here we are now, making our small cameo appearance in the immense cosmic show, pondering the meaning of our presence in the universe.

But let us not forget that it was all started, and is all kept together, by something deeply different from us: the cosmic scaffolding that grew the galaxies we live in and keeps them together is made of a form of matter that is unknown to us, and far more abundant in the universe than any form of matter we have touched, seen, or experienced in any way. It is now time to discuss the strategies so far devised to solve this mystery and to finally identify the nature of dark matter particles.

5

The Swiss dark matter factory

One Ring to rule them all,
One Ring to find them,
One Ring to bring them all
and in the darkness bind them.
J. R. R. Tolkien (1892–1973), *The Lord of the Rings*

A red hydrogen bottle lies in a small underground hall below CERN (the name is derived from "Conseil Européen pour la Recherche Nucléaire", or "European Council for Nuclear Research"), at a depth of about 100 meters. It is from this unlikely—and rather low-tech-looking—object that all the protons[50] that circulate in the most powerful and most famous particle collider in the world, the Large Hadron Collider, spring, before being accelerated to velocities very close to the speed of light.

In this chapter, we will see how the tiny flow of gas from this small red bottle of hydrogen can be turned into incredibly powerful beams of protons, and how these can be precisely arranged into bunches speeding in opposite directions and smashed against each other at the center of particle detectors.

---

[50] As we have seen, the proton is a positively charged particle that, together with the neutron, is a building block of atomic nuclei.

The ring

Let us follow the acceleration of a proton from the dull-looking hydrogen bottle to the highest energies per particle ever produced on the Earth. A hydrogen atom, made of one proton and one surrounding electron, is pulled from the bottle into a basketball-sized object called a duoplasmatron, which strips off the electron, leaving a "naked" proton. From there, the proton undergoes a series of acceleration steps (Figure 5.1) as it jumps from one accelerating machine to the next, like a manual-transmission car shifting gears as it increases its velocity.

The proton shifts into first gear as it enters the three stages of a 30-meter-long linear accelerator, the LINAC 2. Thanks to a a careful arrangement of oscillating magnetic fields, the proton emerges from the LINAC 2 with an energy[51] of 50 MeV. At this

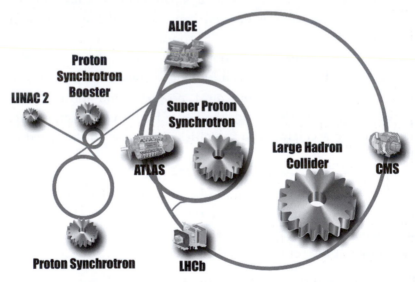

Figure 5.1: A schematic illustration of the Large Hadron Collider five-step acceleration process, and the locations of the main detectors.

---

[51] We refer here to the *kinetic* energy of the proton, that is, the energy associated with its velocity. The proton has also an energy associated with its mass, which is, as we have seen, about 1 GeV.

point, our proton is already traveling at one third of the speed of light. If shot towards the sky, it would reach the Moon in less than four seconds!

The proton is then guided into one of the four rings of the Proton Synchrotron Booster, where it shifts from first to second gear. The booster is a machine built in 1972, about 50 meters in diameter, that still keeps its 1970s look with its orange and green boxes of electrical equipment. Thanks to a combination of pulsed electric and magnetic fields, the proton reaches 90% the speed of light, or an energy of 1.4 GeV. In fact, it does not travel alone: 500 *billion* protons go through the booster every second, and it is here that they are arranged into evenly spaced bunches.

Along with its bunch-mates, our proton shifts gear again as it enters the Proton Synchrotron. When it was built in 1959, this 628-meter-long machine was for a short period of time the most powerful accelerator in the world. But even after many upgrades, the maximum energy the proton can reach in the Proton Synchrotron is "only" 25 GeV, and two much larger accelerators are required before it can reach the energies required to discover new physics.

The first of these machines is the Super Proton Synchrotron, ten times larger than the Proton Synchrotron, and the second largest machine in the CERN complex, and it allows our proton to shift into fourth gear. This seven-kilometer-long accelerator has been of extraordinary importance for particle physics, as it led to the discovery of the W and Z particles, for which Carlo Rubbia and Simon van der Meer were awarded the Nobel Prize in Physics in 1984. When our proton exits from it, its energy is about 450 GeV.

Finally, the proton is injected along with its bunch into one of the two 27-kilometer-long vacuum pipes of the Large Hadron Collider, one hosting proton bunches rotating clockwise and the other hosting bunches rotating anticlockwise. Both pipes are filled with the same number of protons; after shifting gears one last time, these will be accelerated to the same final energy of up to

14 TeV, and eventually led to violently collide inside the particle detectors placed along the circumference of the LHC.

At its nominal maximum energy, with two 7 TeV beams circulating in opposite directions in the ring, the LHC will contain about 2800 bunches in its orbit, with several hundred billion protons in each bunch. As in a car race, the distance between bunches is expressed in units of time: 250 billionths of a second.

A quick search on the Internet allows one to compare the energy of the LHC protons with the macroscopic energy of everyday objects. The energy of a single proton is comparable to that a flying mosquito (concentrated in a single particle!), and the energy of an entire bunch of protons is comparable to that of a medium-sized car at highway speed!

This is already an achievement per se, as the technology and infrastructure required to accelerate protons to these incredible energies are extremely complex, but what's the real purpose of producing these extreme energies, and why are these particles then smashed together inside particle detectors? The answer lies in Einstein's famous formula $E = mc^2$, which is best explained in the words of Einstein himself:[52]

> mass and energy are both but different manifestations of the same thing.

By smashing together high-energy protons, we can convert this energy into new particles with a mass much bigger than that of the original protons. The LHC was built to push the total available energy from about 2 TeV, as achieved by the Tevatron at the Fermi National Accelerator Laboratory in Chicago, to 14 TeV. This is a remarkable achievement that has already allowed the discovery of the famous Higgs boson, as described in more detail below, and may or may not lead to the discovery of other particles.

In fact, when particles collide, they produce a host of secondary particles. The net result of a collision is similar to an explosion that sends debris (the particles created in the collision)

---

[52] A quick Internet search will allow the interested reader to find a recording of Einstein's voice explaining the meaning of this equation.

everywhere. Each collision generates a different combination of particles, moving in different directions, but the laws of physics impose very precise constraints on the type, direction, and energy of the debris.

The most important constraint is that the energy of the debris must be carefully balanced, so that if you see an energetic particle in your detector moving upwards, there must be one or more particles carrying the same amount of energy in the opposite direction. But there are also many other constraints, which regulate the frequency with which a particular pair of particles can be produced, their angular distribution, and their eventual decay into lighter particles.

All these details are encoded in the mathematical formulation of the Standard Model of particle physics, and it is the job of theoretical physicists to make predictions that can be subsequently tested with detectors. Experimentalists have the tough job of building a detector that can keep track of all the particles produced in each collision and precisely reconstruct their nature, direction, and energy.

To this end, big particle detectors have been built specifically for the Large Hadron Collider. When I first visited CERN, I couldn't help noticing the old buildings, the dusty corridors, and the overall rather grimy look of the section hosting the theory institute. But it was when an elevator took me down to visit the accelerator that I realized the colossal size of the detectors, and the incredible degree of sophistication of the technology used.

ATLAS, for instance, is a 25-meter-high, 25-meter-wide, 45-meter-long detector, and weighs about 7000 tons. Someone has called it the "sensitive giant".[53] Its layout is portrayed in Figure 5.2. This detector is optimized to detect all of secondary particles arising from the collision point using an onion-shell structure, comprising the following units:

- The *Pixel Detector* lies very close to the region where the protons collide. It is in effect a small cylinder with a radius between 5 and 15

[53] <http://tinyurl.com/9m4q2uc>.

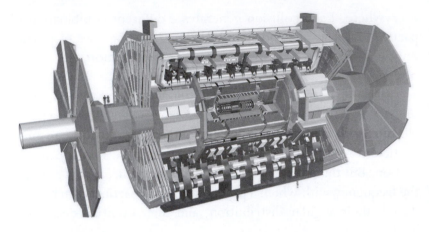

Figure 5.2: The "sensitive giant": CERN's ATLAS detector.

centimeters and a total length of 1.6 meters. The aim of this detector is to precisely track the trajectories of all charged particles produced in the collision. To do so, it is tiled with tiny modules containing a layer of silicon and a layer of electronics, separated by an array of solder spheres. When charged particles pass through the detector, they free some of the electrons in the silicon layer. This process induces an electric current running through the solder spheres, which is then detected by the electronics layer. Like the pixels of a photographic camera, the pixels excited by a charged particle can be used to take a snapshot of the trajectory of the particle.

- The *Transition Radiation Tracker* consists of tubes filled with gas, with a gold-plated wire at the center of each tube. This wire collects the electrons produced by photons originating from the interaction of charged particles with the gas molecules. This again allows one to track the direction and energy of charged particles but also to discriminate between particles producing different amounts of photons. It is, for instance, possible to tell the difference between electrons and pions.

- Further out, the *Electromagnetic Calorimeter* is composed of many accordon-shaped lead and stainless steel elements, bathed in liquid argon.[54] In between, a layer of copper collects the electrons

---

[54] See the next chapter for a discussion of the properties of argon.

produced by the interaction of electromagnetic particles with the absorbers (that is, the lead and stainless steel elements). Showers of particles in the argon liberate electrons that are collected and recorded.

- Next we find the *Hadronic Calorimeter*. As in the Electromagnetic Calorimeter, we find here an absorber, made of steel, but this time interwoven with a scintillator, that is, a material that produces light when hit by charged particles. Light is then collected by light guides, which allows the detector to measure the total energy of the incoming particles. This detector is tailored to measure the total energy of the hadrons produced in collisions of protons.

- Finally, in the outer part of ATLAS we find the *Muon Spectrometer*, which consists of thousands of sensors similar to the tubes filled with gas in the Transition Radiation Tracker, but with a larger diameter. These allow detection of the only particles that can cross the Electromagnetic and Hadronic Calorimeters without being stopped: muons.

The entire structure is immersed in an extremely high magnetic field, which allows the experimenters to identify the nature and energy of particles, since highly energetic particles bend less than low-energy ones, and positively and negatively charged particles bend in opposite directions.

Overall, this instrumentation serves the specific, and extremely difficult, purpose of measuring the direction and energy of all of the particles produced by the protons colliding at its center. This is a daunting task: the proton bunches collide 40 million times every second. Even if collisions are relatively rare, the hundreds of billions of protons in every bunch produce around ten collisions per bunch crossing.

This means that the data acquisition system must keep up with 400 million collisions per second, each generating more than one megabyte of data: in principle the system would need to record about one petabyte, therefore filling a thousand 1 terabyte hard disks, every second! Since this is impossible to handle in practice, the ATLAS scientists automatically select the most interesting events on the basis of carefully designed algorithms and discard all

Figure 5.3: As many as 25 collisions can be detected in this extreme case of pileup in the CMS detector, which complicates the reconstruction of the nature and energy of the debris arising from individual collisions.

other events, reducing the data load to "only" 15 petabytes of raw data per year.

There are several other detectors around the orbit of the LHC, six in total. ATLAS actually has a fraternal twin, CMS (Figure 5.3), which is similar in scope to ATLAS but different in structure. These are the two factotums of physics, designed as two giant multipurpose experiments that will be able to shed light on the widest possible range of physics searches.

It is crucial to have two independent experiments: as an old Roman legal adage goes,

Testis unus, tests nullus,

which translates approximately as "a single witness is as good as none". This is a worry that will manifest itself in all its virulence in the next chapter, when we discuss the conflicting results of the many underground experiments searching for dark matter.

The other detectors are ALICE, an experiment that aims to re-create a form of matter, the quark–gluon plasma discussed in the previous chapter, that existed in the first instants of the universe; LHCb, which aims to study the asymmetry between matter and antimatter in the universe; TOTEM, which will provide a measure of the size of the proton, among other things; and LHCf, which will provide useful information for the community of scientists studying cosmic rays.

The more pragmatic readers may wonder what the cost of all this is. If we consider just the machine, excluding the cost of personnel, we get a cost of about 3 billion euros. The superconducting magnets are responsible for about half of the cost. Another quarter arises from civil engineering works and the cryogenic system that is needed to cool the machine, and the remaining quarter from the ventilation system, power distribution system, and so on. In the last chapter of the book, we will come back to the costs of particle physics and its social and economic benefits.

Into the unknown

As we have seen in the previous chapter, there are reasons to believe that the Standard Model of particle physics represents only a subset of all the particles and forces that exist in nature, and there are well-motivated extensions of the Standard Model that predict the existence of new particles that turn out to be perfect dark matter candidates.

The first *weakly interacting particle* proposed as an explanation for dark matter was the Standard Model *neutrino*. This idea was put forward in 1973 in a famous paper written by two scientists at the University of California, Berkeley: Ram Cowsik (who was on leave from the Tata Institute for Fundamental Research in Mumbai, India) and J. McClelland. One week before submitting this paper to *Astrophysical Journal*, these two authors had submitted another paper to *Physical Review Letters*, in which they used cosmological considerations to set an upper limit on the mass of the neutrino. This was already interesting per se, but in the second paper they took a much bolder step, by *identifying* neutrinos with dark matter and studying the astrophysical consequences of this hypothesis.

It probably took the two young scientists a lot of self-confidence to submit such a radical proposal to a prestigious international journal, at a time when dark matter was thought by many to be in the form of stars or gas, or simply not to exist at all. But a sentence in their paper reveals that they were not the only ones to have thought about this possibility:[55]

> Though the idea is undoubtedly not new, it does not appear to have been presented in published form before.

A giant of astrophysics, the Russian Zel'dovich, constructed an elegant theory in the late 1970s and early 1980s that explained how astrophysical structures could form under the influence of dark matter made of neutrinos, but it wasn't long before this model was ruled out: in 1984, computer simulations run by Simon White, Carlos Frenk, and M. Davis demonstrated that the neutrino hypothesis was incompatible with observations.

Meanwhile, in 1977, Benjamin Lee and Steven Weinberg had suggested the possible existence of *heavy leptons* and used cosmological arguments to constrain their mass. It took only a few years before this argument was applied to particles arising from a theoretical framework that was already emerging back then as one of the most promising ways to address the problems of the Standard Model: supersymmetry. Several groups of authors suggested this connection. Pagels and Primack proposed the gravitino—the supersymmetric partner of the graviton—as a promising dark matter candidate. Goldberg, in 1983, and then Ellis, Hagelin, Nanopoulos, Olive, and Srednicki, in 1984, argued that one should instead search for the dark matter particle among the partners of the Higgs and gauge bosons, today called neutralinos.

Interestingly, whereas the possible *identification* of the gravitino with dark matter was explicitly mentioned in Pagels and Primack's paper, the papers focusing on the neutralino simply used cosmological observations to constrain its properties. (The lightest of the neutralinos is the only one that is stable, and therefore the only one that is a viable dark matter candidate; it is often referred to as *the* neutralino.)

---

[55] R. Cowsik and J. McClelland, "An upper limit on the neutrino rest mass", *Physical Review Letters* 29 (1972) 669.

The supersymmetric neutralino is a prototypical example of a WIMP (weakly interacting particle). Its mass can range from about 50 to a few thousand times the mass of the proton and its interactions with ordinary matter and with itself are such that it can account for all the dark matter in the universe while still remaining consistent with all known experiments.

It is not uncommon to see the two names "WIMP" and "neutralino" used as synonyms, but this is inappropriate: it is important to keep in mind that WIMPs might not be neutralinos, as we have seen in Chapter 3, and that dark matter might not be made of WIMPs. To make things more complicated, the theory of supersymmetry is, in its general form, almost completely unconstrained: it has a large number of *free parameters*—quantities that regulate, among other things, the properties of the neutralino, including its abundance in the universe and its interactions with other particles—and none of the new particles it predicts have been discovered yet.

How can we search for neutralinos, or for WIMPs in general, at the LHC, if we know so little about their properties? The answer for the case of a generic WIMP is simple: we can't. We will see in the following chapters that the situation is different for direct and indirect searches, for which a general strategy can be devised even in the absence of a detailed understanding of the underlying theory. But accelerator searches require very precise information about the *phenomenology* of particles—their production rate in proton–proton collisions, their decay time, the type of secondary particles they produce when they decay, and so on.

In the case of neutralinos, as with the particles arising from other extensions of the Standard Model proposed in the last 30 years, detailed predictions can be made of all of these properties. It is by comparing predictions with actual data that theoretical and experimental physicists can together discover new particles or rule out wrong hypotheses.

To understand how this is done in practice, let us go back to the schematic description of the possible structure of theories that seek to extend the Standard Model (see Figure 3.3). This description is valid for supersymmetry and also for several other models, including theories that predict the existence of additional space dimensions.[56]

Each new particle can in principle decay into lighter particles. Suppose for instance that when two protons are smashed together, the quarks inside them collide to produce their supersymmetric partners—that is, *squarks*. Once these squarks are created, they immediately start what is called a decay chain, a series of reactions in which they decay into lighter and lighter particles.

One of the rules of the decays in the most common extensions of the Standard Model, for instance supersymmetry, is that each time a particle of the new theory decays, its decay products must contain at least one other particle of the new theory. In most supersymmetric theories, the neutralino is the lightest new particle. By virtue of this rule, it cannot decay into any lighter particle, and it is therefore *stable*: all the supersymmetric particles produced in the early universe decay into neutralinos, and these survive.

Let us see how this could work in practice. Figure 5.4 shows a possible decay chain arising from a high-energy proton–proton collision at the Large Hadron Collider. First, a massive squark is produced. But the squark is unstable and decays into two particles. As one can see from the figure, the decay chain includes a particle of the Standard Model, a quark, and another supersymmetric particle, the *second* neutralino. This in turn decays into a slepton—the supersymmetric partner of a lepton—and a Standard Model lepton. Finally, the slepton decays to the lightest neutralino and another lepton.

If a decay chain like the one depicted in Figure 5.4 develops inside ATLAS or CMS, it will—at least in principle—be easy to detect it. The two leptons and the quark will leave unmistakable signatures in the detector, but when scientists try to reconstruct the total energy of the event, they will find an imbalance in the

---

[56] See pages 57–59, under "Terra incognita" in Chapter 3, for more details of theories that predict the existence of extra dimensions.

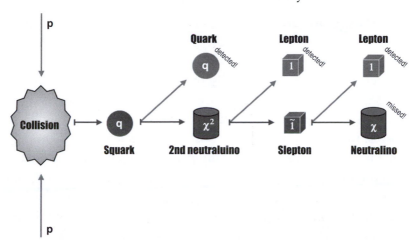

Figure 5.4: A possible decay chain from a heavy particle produced in a high-energy collision at the LHC to the dark matter particle.

total energy:[57] the "smoking gun", in the case of a perfect detector, of the production of new particles that have escaped from the collision point without leaving a trace.

The problem is that since the effect one is searching for is very small, even the smallest imperfection in the detector—"dead pixels", mismeasurement of energy, inactive material, and so on—may fake missing energy or degrade the precision of the measurement. The experimental collaborations have shown convincingly that it is possible to keep all these uncertainties under control and to perform measurements with exquisite precision at both ATLAS and CMS. An extraordinary demonstration of the quality of the detectors came from a discovery that has been celebrated as the most important physics result of the century: the discovery of the Higgs boson.

---

[57] Usually, only the energy emitted in a direction perpendicular to the beam direction is taken into account. The reason is that the collision involves the subnuclear components of protons, namely quarks, and gluons, which carry an unknown fraction of the proton energy. The total energy in the beam direction can therefore exhibit an imbalance even in the absence of new particles escaping detection.

The Higgs boson

On the fourth of July, 2012, all particle physicists on the planet rejoiced at the extraordinary discovery of a new particle with the ATLAS and CMS detector (Figure 5.5). Not only had a new boson been discovered, but its properties also looked very much like those of the long-awaited Higgs boson, a particle of central importance for particle physics. For my generation, this was by far the most important physics discovery that we have ever witnessed, and perhaps the same is true for more senior physicists also. The reason is that this particle allows us to validate the remaining missing piece of the Standard Model puzzle: the origin of particle masses.

    Much has been said in the media about the Higgs, and there are excellent sources of information on the Web for those who want to understand the importance of this discovery. Let us summarize the key points here.

Figure 5.5: An excited crowd of scientists applauds Peter Higgs (second standing from the left) and codiscoverers of the Higgs mechanism on the occasion of the announcement of the discovery of the Higgs boson on July 4, 2012, in the CERN auditorium.

First, it is not the discovery of the particle per se that is important, but the fact that it allows us to validate what is called the *Higgs mechanism*, which postulates the existence of an entity, the Higgs field, with which all massive particles interact.

Second, the Standard Model was flawed until July 4, 2012: the most elegant theory of particles and their interactions predicted the existence of massless particles, in obvious contradiction with the reality of the world we live in. The Higgs mechanism, invented in 1964 by Peter Higgs and simultaneously by two teams of scientists—one including Robert Brout and François Englert, and the other including Gerald Guralnik, C. Richard Hagen, and Tom Kibble[58]—was known to provide a much-needed cure for this disease of the theory, but all previous attempts to discover the Higgs particle had failed, and doubts had started to creep into the community as experiments like LEP at CERN and then the Tevatron at Fermilab excluded a larger and larger range of possible masses.

Third, we are not sure yet whether the particle discovered at CERN is indeed the Higgs boson or just a particle that closely resembles it. Future data will allow us to perform some crucial tests that should convince even the most skeptical scientists.

Let us turn now to the implications of the discovery. The Higgs mechanism postulates that particles acquire mass as they travel through the "sea" of the Higgs field. In fact, the Higgs field is ubiquitous—and we'd better start to get used to this idea now that there is convincing proof of it—and particles which would be massless in the absence of the Higgs field interact with the field itself. This interaction slows down particles, which then behave effectively as if they had a mass.

The mass of a particle is therefore determined by the strength of its interaction with the Higgs field. The fact that the electron is lighter than, say, the top quark is due to the fact that the top quark interacts much more strongly with the Higgs field than does the electron. Similarly to the electric charge, which encodes the

---

[58] The interested reader will find a historical account at <http://arxiv.org/abs/hep-th/9802142>.

strength of the electromagnetic interactions of a particle, the inertial mass is a different type of "charge" that measures the coupling of a particle with the Higgs field.

How did CERN scientists manage to discover the Higgs boson among the many secondary particles produced in each collision? From the mathematical formulation of the Higgs mechanism, one can calculate the probabilities for such a particle to decay into different channels. One of the most simple and "clean" is a decay into two photons with identical energies, as shown in the case of a real event recorded by the CMS collaboration in Figure 5.6.

Once the experimental collaborations have smashed together enough protons, they can then look at the data and count the number of events in which two photons with identical energies have been produced. Figure 5.7 shows an adapted version of the actual

Figure 5.6: A real event measured by the CMS detector at CERN, possibly due to the decay of a Higgs boson into two photons (whose energy is shown by the diagonal streak and the dots in the bottom left corner). The tracks of all particles produced in the collision are also shown.

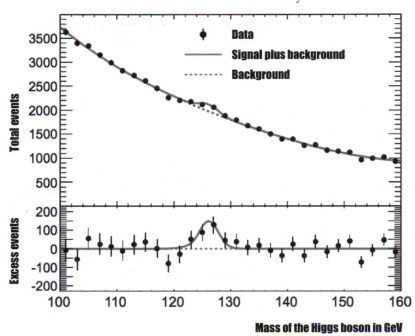

Figure 5.7: The discovery of the Higgs.

data taken by the ATLAS collaboration:[59] by counting the events at different energies, shown as filled circles in the figure, the collaboration could search for an excess of these events (fitted with a solid curve in the figure) on top of the background (dashed curve) produced by the usual Standard Model processes.

This excess was in fact measured in the two-photon channel, and also, by an independent analysis, in other channels. Even more importantly, it was found at the same mass, 126 GeV, by the two experimental collaborations *independently*: a guarantee that the observed excess is not due to calibration problems or other experimental errors.

---

[59] This figure is adapted from an analogous one that appeared in the paper in which the ATLAS collaboration announced the discovery of the Higgs boson; see the list of figures at the end of the book for the full reference.

As important as the discovery of the Higgs boson is, much is still to be understood in particle physics. The properties of the newly discovered particle need to be studied in great detail, in order to prove that it actually corresponds to *the* Higgs boson, the one now incorporated into the Standard Model. As more data become available and we get a sharper view of the long-sought particle, we can check whether it behaves exactly as predicted or as a different "monster", perhaps with properties similar to one predicted by new physics theories such as supersymmetry.

The unspoken fear

Until we run the LHC at its maximum energy, we will not know whether supersymmetric particles, or anything else beyond the Standard Model, exist. It is possible that our theoretical description of fundamental particles may be flawed, or that new particles do exist but they are so massive that they cannot be produced at the LHC. That's the unspoken fear of particle physicists today: that the Standard Model is all there is, or at least all that can be found with our accelerators.

Certainly, many physicists will have reason to be disappointed in that case. Some of the most prominent theoretical physicists alive have put decades of work into the study of theories beyond the Standard Model, and it would obviously hurt to be forced to abandon the hope of seeing them experimentally verified within their lifetime. But, personally, I wouldn't consider that as a failure of the community as a whole: the timescale of experiments has grown exponentially from the tabletop experiments of the beginning of the 20th century, and it takes a great deal of time, money, and international coordination to build the powerful accelerators that can put such theories under test.

Take the LHC. Its construction was proposed in the 1980s and was approved in 1994. But it wasn't until 2009 that the actual scientific program could finally start. It is inevitable that the very nature of scientific research will be affected by the longer and longer timescales required to build even larger machines, and it is inevitable that theoretical research will go in the direction of making increasingly detailed predictions rather than making up new theories that cannot be tested with experiments.

The next step depends crucially on what is found by the LHC after it doubles its energy in 2014, after a long shutdown phase in 2013 to allow technical updates. If new particles are not discovered within the first two years, then the hopes of finding new physics will rapidly vanish. It would be difficult at that point to justify the construction of an even more powerful accelerator: if new physics does not exist at the TeV scale, where it could help in explaining the difficulties of the Standard Model, then it could be anywhere.

The other possible outcome is of course that new particles *are* actually discovered: a circumstance that would change particle physics forever, and motivate large investments in new, more powerful machines to allow us to discover what lies on the "dark side" of the universe.

Particle physicists are, meanwhile, thinking about the next machine. One possibility is to reuse the LHC tunnel to construct LEP3—a new electron–positron accelerator, like LEP and LEP2, which were operated in the same tunnel before the LHC was constructed. The problem with this option is that the energy achieved by such a machine would be limited, probably making it difficult to discover any new physics beyond the Standard Model. This is due to the fact that electrons lose energy as they move in a curved trajectory, and a tunnel even bigger than the current 27-kilometer-long one would be needed in order to achieve energies well above the TeV range. Although the hypothesis of digging an 80-kilometer-long tunnel has been discussed at CERN, it is unlikely that this will happen anytime soon.

An interesting alternative that appeals to many scientists is a *linear* electron–positron collider. The most widely discussed designs consist of a 30-kilometer collider reaching an energy of 1 TeV—a project known as the ILC, for "International Linear Collider"— and a 50-kilometer version that aims to reach up to 3 TeV thanks to a novel technology for the acceleration of particles—called CLIC, for "Compact Linear Collider". The cost of such experiments is, however, very high; they can probably be built only in the framework of a worldwide collaboration, possibly led by Japan according to insistent rumors, and only if a strong physics case can be made for their construction after many years of LHC data have been accumulated.

Finally, there is strong interest in the community in a circular *muon* collider, which could achieve multi-TeV energies with a very compact design compared with an electron–positron collider. However, many technological hurdles need to be overcome, including the extraordinary challenge of *cooling* the muons (that is, arranging them into a narrow beam) and accelerating them within 2.2 millionths of a second, before they decay into electrons and neutrinos.

As the LHC accumulates data and grand plans are made for the future of particle physics, a new generation of astroparticle experiments is about to start taking data. They are expected to provide complementary information about possible extensions of the Standard Model, and they may even allow us to discover dark matter particles before they are found in accelerators. In the next two chapters, we will see how.

6

Underground mines, with a view

People in pits and wells sometimes see the stars.
Aristotle (384 BC–322 BC), *On the Generation of Animals*

According to an old legend, it is possible to see stars in the sky even in full daylight, provided that one observes them from a deep enough well. It is obviously a false belief, but one with a truly venerable history. It can be traced back to the writings of Aristotle[60] and, quite surprisingly, it survived until very recently, even in some of the most famous literary works, such as the well-known *Pickwick Papers* by Charles Dickens,[61] despite the fact that it can, and has been, easily disproven.

Still, some of the most advanced physics laboratories for the study of the universe are located in deep mines or in the heart of large mountains. The explosion of a nearby supernova in 1987, for instance, was observed with a neutrino telescope located underground, 1000 meters under Mount Kamioka in the Gifu Prefecture of Japan. Why?

To answer this question, let us take a look at what happens inside one of the most important laboratories for astroparticle physics: the Gran Sasso laboratory. The name of the lab comes from the Gran Sasso d'Italia mountain, the highest in Italy outside

---

[60] See, e.g., D. W. Hughes, "On seeing stars especially up chimneys", *Quarterly Journal of the Royal Astronomical Society* 24(3) (1983) 246.

[61] Opening of Paper 20.

the Alps, and one of extraordinary beauty. A tunnel was constructed through it in the 1960s to allow faster travel between the western and eastern coasts of Italy, and in 1982 the physicist Antonino Zichichi had the idea of excavating a hall near the center of the tunnel to host a physics laboratory.

The idea turned out to be a great one, and the laboratory currently hosts 15 experiments, including some of the most important for the search for dark matter. The common aspect of all of the experiments placed underground is that they are searching for *rare events*. In Chapter 1, we used the metaphor of a person receiving a phone call in a noisy environment such as a busy restaurant to illustrate the need to move to a quiet place when dealing with weak signals.

If one is trying to find direct evidence for dark matter particles, underground is the place to go. The reason is that dark matter interacts very, *very* weakly with ordinary matter. If you placed a detector on the surface of the Earth or in a shallow location, all you would observe would be cosmic radiation: particles showering down from interstellar space and hitting the nuclei of your detector. The result is that the rare collisions between dark matter particles and the detector nuclei would be buried in a sea of collisions of other types, therefore making the identification of dark matter impossible.

Shielding against cosmic radiation requires kilometers of dense rock, and therefore physics experiments searching for rare events are placed deep underground, where only a very small fraction of cosmic rays penetrate, and one can "listen" better to the hits produced by collisions of dark matter. Rock acts in a sense like a filter: cosmic radiation is absorbed, but dark matter travels undisturbed through it.

The problem now is how to detect particles that can travel undisturbed through kilometers of rock: the probability that they will interact with an experiment placed underground must be extremely small! That's true, but the smallness of the interaction is compensated for by the large number of particles traveling through the detector every second: each of them has only a small chance of hitting a nucleus in the detector, but if at least one hits the detector in, say, a year, then you can try to catch the tiny amount of energy that it deposits in your experiment.

In order to do that, we need to make an experimental apparatus that is big enough to give us a large probability of interaction; this probability is proportional to the number of nuclei that can be hit by dark matter particles, and therefore to its mass. We need also to place the apparatus in an environment where there are few particles that can mimic a dark matter interaction, which implies placing it underground and then shielding it from the particles produced by the natural radioactivity of the rock and other materials.

The last three decades have seen an impressive improvement in the sensitivity of dark matter detectors. In this chapter, we will take a look at the most interesting technologies devised so far to detect these elusive particles, and at the prospects for detecting dark matter in the upcoming generation of experiments.

Ups and downs

H. G. Wells, in his *Short Stories*, beautifully described the solitary journey of the solar system in interstellar space:

> The sun with its specks of planets, its dust of planetoids, and its impalpable comets, swims in a vacant immensity that almost defeats the imagination.

We have learnt a lot in the last three decades about this "vacant immensity". We know in particular that it is not really vacant, and that the solar system cruises instead through large amounts of dark matter, at a speed of approximately 220 kilometers per second, following a circular orbit around the center of the Galaxy. To an observer on the Earth, it is as if these particles were "raining" on the solar system from a very specific point in the sky, in the constellation Cygnus, corresponding to the direction towards which the Sun moves along with its "specks of planets", including the Earth.

As the Earth revolves around the Sun, it moves against the direction of the raining particles in June, and in the same direction in December (Figure 6.1). Just as the number of raindrops hitting the windshield of a car when it is raining is larger when one is driving upwind, i.e., when the velocities of the car and the wind are in opposite directions, the number of dark matter particles hitting an underground experiment is larger when the velocity of the Earth is aligned with the motion of the Sun, and is therefore opposite to the velocity of the dark matter rain.

Based on this argument, Andrew Drukier, Katie Freese, and David Spergel noted in 1986 that an experiment able to detect dark matter particles would measure a higher number of particles in summer than in winter. This idea was prompted by the seminal work of Drukier and Leo Stodolski in 1984, who proposed a technique to detect neutrinos using their interactions with the nuclei of a detector, which was subsequently generalized by

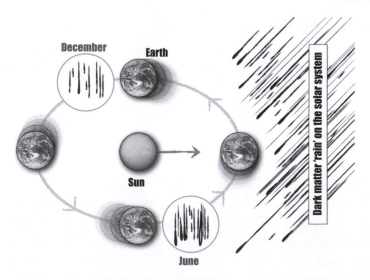

Figure 6.1: Owing to the motion of the Earth around the Sun, dark matter particles from the Milky Way halo rain on the Earth in June with a stronger intensity than in December.

Mark Goodman and Ed Witten in 1985 to a variety of dark matter candidates.

After years of preliminary results, the DAMA/Libra collaboration announced in 2008 that this effect had been measured beyond reasonable doubt. To be more specific, the experimental collaboration published a paper claiming that "the presence of dark matter particles in the galactic halo is supported at 8.2σ".

The "number of σ" is a measure of the probability that the experiment could have obtained data supporting an annual modulation *by chance*, owing to a statistical fluctuation in the data, in the absence of an actual physical effect. The higher the number of σ, the smaller the probability that this could have happened. By convention, particle physicists speak of *evidence* for new particles when the probability is about one in 740, corresponding to 3σ, and of *discovery* when the probability is about one in 3.5 million, corresponding to 5σ.

It may sound a bit involved, but a quick example will perhaps elucidate the meaning of this convention better. Suppose that you toss a coin and that it always falls on the same face. After a few tosses with identical results, you may legitimately start to be suspicious, but after ten tosses, you are entitled to claim evidence of treachery, and after 22, you can finally claim discovery of fraud (Figure 6.2). The 8.2σ result announced by the DAMA

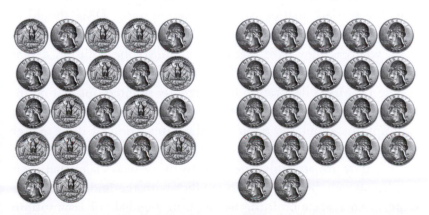

Figure 6.2: *Left*: random sequence of coin flips. *Right*: 5-sigma evidence of fraud.

collaboration corresponds to a probability of one in *ten million billion*, which is that of throwing all heads, or all tails, 54 times in a row, which is perhaps overkill.

Now the obvious question is: could the modulation of the event rate have been induced by dark matter particles? All conventional explanations, from the obvious to the most subtle, have been ruled out by the DAMA/Libra collaboration, yet the community remains skeptical, a circumstance that triggers lively discussions at physics conferences dedicated to dark matter.

The debate has risen to unusual heights for the community of physicists, which traditionally prides itself on openness, objectivity, and rationality. This, for instance, is a quote from Kipling's famous poem *If* that appears on the website of the DAMA/Libra experiment:

> If you can bear to hear the truth you've spoken
> twisted by knaves to make a trap for fools,
> you'll be a Man my son.

Who are the "knaves" twisting the truth of the DAMA/Libra experimentalists? Who are the "fools" deceived by those "knaves"? As we have seen, many other experiments are currently searching for dark matter. Some of them, like the Xenon and CDMS experiments described below, are supposed to have already achieved a much better sensitivity than DAMA. If the signal found by DAMA was due to dark matter particles, Xenon and CDMS should in principle have seen tens or hundreds of collisions in their detectors, whereas they have measured at most one collision a year or so.

The DAMA collaboration has correctly pointed out that no direct comparison of their result can be made without making some assumptions about the nature of the interaction of dark matter with the detector nuclei. When they refer to "knaves twisting the truth", they are arguably referring to those who claim that their result is wrong because it is in conflict with the findings of other experiments while omitting to mention the chain of assumptions made to reach this conclusion.

These hidden assumptions include the velocity of dark matter particles in the Milky Way halo, the details of the interaction of dark matter with the nucleons in the detector, and the properties of nucleons inside a nucleus. Each of these quantities has its own uncertainties, and theoretical physicists have therefore tried to see whether making different choices for them could help reconcile all of the experimental results.

Although many attempts have been and are being made, this seems a very difficult task. Also, to make things more complicated, the experimental collaborations CoGeNT, CRESST and CDMS II, have measured a rate of events in their detector which is substantially higher than the expected background, but the dark matter parameters required to explain this rate are not exactly the same as those required to explain the DAMA data.

The difficulty of reconciling these results with the other (null) experimental searches leaves only two options open: either the standard description of dark matter particles is flawed or at least one experimental result, or its interpretation, is wrong. In principle, at least in the case of DAMA, the issue could be easily solved by repeating the experiment somewhere else, to ensure that the signal does not arise from instrumental or environmental effects. It is, however, hard to convince an experimental group to invest its energies in such an endeavor, since it would imply investing in an "old-generation" experiment, with the possible outcomes of confirming a result found first by DAMA, which they would not be in a condition to claim ownership of, or ruling it out, therefore confirming what many other experimental collaborations are currently suggesting. Everyone would like such an experiment to be built, just by someone else.

Fortunately, someone has recently decided to take up the challenge. An experiment that is being constructed at the South Pole, called DMIce, aims to use crystals of the same type as used by DAMA, sodium iodide, to perform a dark matter search in the southern hemisphere. Since the experiment would be very similar to DAMA, it should see an identical signal, with its characteristic annual modulation, if DAMA's signal arises from dark matter interactions, whereas the signal should be different or disappear if that is not the case.

To achieve the extraordinary quality of the crystals achieved by the DAMA collaboration is complicated, however, and the construction of the experiment will likely require several years of dedicated effort. While the DAMA result is being checked, a large number of other experimental collaborations have started an impressive race to discover dark matter, adopting a variety of experimental strategies. We will now take a look at some of the most promising ones.

The German

In 1985 Blas Cabrera, Lawrence M. Krauss, and Frank Wilczek came up with an idea to measure neutrinos using a *bolometric* technique: they demonstrated that the small amount of energy deposited by a neutrino interacting with a detector would induce a small but measurable increase in the temperature of the detector.

It didn't take long before Cabrera and Bernard Sadoulet applied this technique to the search for dark matter particles, combining it with a clever and technologically challenging technique to distinguish the energy deposited by dark matter particles from most of the interactions of ordinary particles in the detector. They chose *germanium* crystals as the sensitive material in their detectors—an element whose Latin name, evidently meaning "the German", was chosen by its discoverer Clemens Winkler in honor of his home country, and which back in 1886 provided a beautiful confirmation of the ideas that had led the famous chemist Mendeleev to the construction of the periodic table.

The choice of germanium was motivated by scientific arguments; for instance, the high atomic number makes it a target that is easier for dark matter particles passing through it to strike. There was also the practical reason that strong expertise existed "in house" in the physics department at Berkeley when Cabrera

and Sadoulet set out their research program. The final design of
the experiment, which was called Cryogenic Dark Matter Search,
or CDMS, is representative of a broad class of experiments,
generically called *cryogenic* experiments, as they need to be cooled
down close to absolute zero.[62]

The energy deposited by dark matter, or any other particle, in
the detector is measured by two methods. The first is by measuring the vibrations of the crystal lattice induced by particles striking
the nuclei of the crystal. Imagine that you have hit the rail of a
balcony with a hammer: the vibrations propagate through the entire railing and can be felt by anyone with his or her hands on it.
When a particle strikes a nucleus in the lattice structure of a crystal, something similar happens: a special type of vibration, called
phonons, propagates in the crystal, excited by the recoil energy of
the nucleus struck by the dark matter (Figure 6.3).

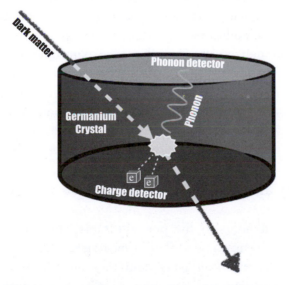

Figure 6.3: Cryogenic experiments like CDMS and EDELWEISS attempt to detect
dark matter particles by measuring the phonon and charge signals produced in
collisions with detector nuclei.

---

[62] More precisely, these experiments are cooled down to millikelvin
temperatures.

The energy of these phonons is tiny. In order to detect it, the experimentalists cool the detector crystal down to temperatures close to absolute zero, and stick a special sensor to its surface that exploits a physical effect known as superconductivity to measure the small increase in temperature induced in it by a phonon.

Feeling with our hands the vibrations induced by an object hitting a rail would not, however, tell us much about *what* had hit the rail. But perhaps we could try to collect additional information. If we knew, for instance, that there were only two possibilities, say a metal hammer or a mallet, and we knew for sure that the metal hammer produced sparks when it hit the rail, we could look (even in complete darkness) for whether sparks were produced every time we felt a vibration, and from that observation we could distinguish hammer hits from mallet hits.

In cryogenic experiments, something similar happens. In addition to the phonon signal, cryogenic experiments also measure an *ionization* signal, by collecting electrons stripped from atoms during particle collisions, or a *scintillation signal*, that is, the light produced in such events. This allows the experimenters to identify signals that might be due to dark matter particles, which produce very little ionization or scintillation signal, from the much more numerous and much less interesting signals induced by ordinary particles.

The Lazy and the Alien

Of all the elements in the periodic table, those occupying the last column occupy a special role in dark matter searches. They are known as the *noble gases*, and they are characterized by a very low reactivity and by being odorless and colorless. There are many good reasons why noble gases are used in dark matter searches: they produce a relatively large amount of light and electric charge when a particle interacts with them and they are relatively cheap, so experiments can be scaled up significantly at a relatively low cost.

This is how Primo Levi introduces them in his famous book *The Periodic System*:

> There are the so-called inert gases in the air we breathe. They bear curious Greek names of erudite derivation which mean "the New," "the Hidden," "the Lazy," and "the Alien." They are indeed so inert, so satisfied with their condition, that they do not interfere in any chemical reaction, do not combine with any other element, and for precisely this reason they remained undetected for centuries.

The "lazy" one is argon, a gas that is relatively abundant in the Earth's atmosphere: only 22 times less abundant than oxygen, and 23 times more abundant than carbon dioxide. Most likely, there are small containers of argon very close to you as you read these lines: incandescent lightbulbs are in fact filled with a special inert atmosphere mostly made of argon, to prevent oxidation of the filament.

But atmospheric argon is not good enough for dark matter experiments. It contains some argon-39—an *isotope* of argon with 39 protons and neutrons in its nucleus, instead of 40 like the substance known simply as "argon". This is a pernicious gas that is produced by the interaction of cosmic rays with the Earth's atmosphere, and undergoes radioactive decay about 269 days after its formation, inducing an unwanted signal in the detector that spoils its sensitivity to dark matter particles.

The researchers of the DarkSide collaboration, one of the leading experimental collaborations using argon, are therefore extracting this gas from pockets of gas buried deep underground in southwestern Colorado. The reason to go to this remote location is that industrial facilities here exist there to extract CO_2, a gas sadly known for its greenhouse effect. This gas is then pumped through a 1000-mile-long pipeline to the oilfields of west Texas, where it is injected into oil reservoirs to increase the amount of crude oil that can be extracted from them.

The CO_2 extracted from the ground contains a small amount of *pure* argon, not contaminated by argon-39. An extraction plant exists in Colorado that separates the argon from the CO_2, but to achieve the purity required by the DarkSide operation, the gas is then taken to Fermilab, where it is treated before being shipped to its final destination: the Gran Sasso laboratory in Italy.

DarkSide promises to become one of the leading experiments in direct dark matter searches, but argon-based experiments so far have made progress more slowly than other technologies. Another noble gas, however, has already allowed us to dramatically improve the sensitivity to dark matter particles: xenon—"the Alien", for Levi. Much less abundant than argon in the Earth's atmosphere, it has been used extensively in a variety of applications since its discovery in 1898, from anesthesia to nuclear magnetic resonance, and from car headlights to, well, dark matter detectors.

Xenon and argon experiments, just like cryogenic experiments, aim to measure the tiny amount of energy deposited by dark matter particles when they interact with the nuclei of the detector. A phonon signal is not available in these experiments, since they use a target in liquid and gaseous form, in which phonons do not develop. Therefore, these experiments make use of a combination of other techniques.

One of the most sensitive experiments currently in operation, Xenon100, detects electrons and light produced by the interaction of dark matter with xenon nuclei. To measure the number of electrons, the electrons are extracted by means of an electric field from the liquid to a thin layer of gas, where they emit additional light, as illustrated in Figure 6.4—a technique that was pioneered in the mid 1990s by Pio Picchi, Hanguo Wang, and David Cline. By studying the relative size of the light pulses produced at the interaction site and in the layer of gas, it is possible to tell whether a dark matter particle collided with one of the xenon nuclei or whether the signal was produced by a collision that involved, say, one of the electrons surrounding the nucleus.

This "noble liquid" technology is evolving at an impressive rate. In 2005, these experiments could detect a rate of events equal to one dark matter particle interacting with a kilogram of detector a day. Today, the minimum detectable rate is one dark matter

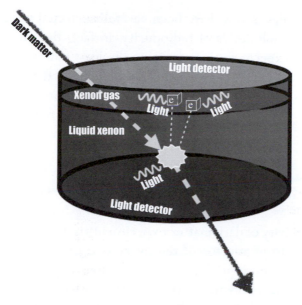

Figure 6.4: The Xenon100 experiment measures the light produced by interactions of dark matter and that produced by electrons that propagate from the interaction point to the layer of gas on top of the liquid xenon.

interaction per kilogram of detector *per year*, and within five years, it will be one event *per tonne* of detector per year.

As in the case of cryogenic experiments, besides the technological challenges of constructing the detectors, the biggest obstacle is the "background". If particles produced by radioactive decays in the rock surrounding the detector or by the interaction of cosmic rays penetrate into the detector, they can wash out any dark matter signal, so it is crucial to shield the experiment not only by bringing it down into a mine or under a mountain, but also by constructing shields of suitable material to absorb any dangerous particles and by carefully selecting the materials used for the construction of the experiment.

The Xenon1T experiment, for instance, will be suspended inside a ten-meter-diameter water tank in order to shield it from the ambient radioactivity. All materials used in the experiment, from the detectors to the cryostat, and from the supporting structure to the smallest screw, have been carefully selected to achieve the highest possible levels of radiopurity. In fact, the aim is that the Xenon1T experiment is to become the least radioactive environment on the Earth. If a scientist inadvertently left a banana in the apparatus, the radioactive decay of the potassium contained in it would significantly degrade the performance of the experiment!

Do or die

The sensitivity of dark matter experiments is improving at the impressive rate of a factor of ten every two or three years as new technologies become available and various experimental collaborations join forces to build ever-bigger detectors. Despite our efforts, however, there is to date no fully convincing proof that dark matter particles have been detected in any laboratory experiment: the most important result obtained so far is that we know what dark matter is *not*.

Direct detection experiments in fact allow us to exclude the existence of particles with given properties. The logic of the procedure is simple: one calculates the rate of interactions that a particle with a given mass and a given strength of interaction would produce in a detector, and if the predicted rate is higher than what is measured, then the existence of a particle with such properties is ruled out. This information is encoded in Figure 6.5, which illustrates the current status of direct dark matter searches. The combinations of dark matter mass and strength of interaction in the large shaded areas are ruled out by the CDMS II and Xenon100 experiments. The small shaded areas correspond to the claims of detection made by the DAMA, CRESST, CoGeNT and CDMS II collaborations discussed earlier in this chapter, which clearly fall in a region excluded by other experiments.

As the sensitivity of cryogenic and noble liquid experiments has improved, it has become obvious that it is hard to reconcile the DAMA and CRESST claims with each other and with all other

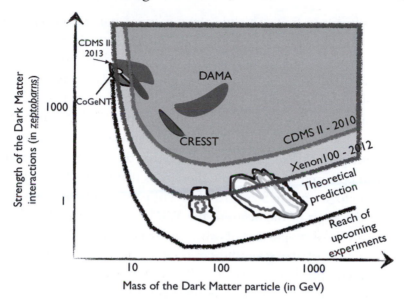

Figure 6.5: The status of direct dark matter searches.

experiments. So, either these claims are wrong or the interpretation that we are giving to these pieces of experimental evidence is flawed.

Figure 6.5 also shows the reach of upcoming experiments, like the successor of Xenon100, Xenon1T, which should become available within a few years, as well as the predicted "most probable" region where dark matter particles should be in the plane of mass versus strength of interaction. This theoretical prediction arises from a particular subset of theories that seek to expand the Standard Model of particle physics;[63] it is the most probable region for the parameters of dark matter particles based on our current knowledge of particle physics and cosmology.

These regions are changing in shape and position as our knowledge of physics improves, usually making the dark matter particle

---

[63] For the experts, we are referring to the constrained Minimal Supersymmetric Standard Model.

more difficult to find, which is obvious, since part of our knowledge is that we have not found it yet. But a consensus is growing that if tonne-scale experiments fail to discover dark matter particles and if there is no sign of new physics at accelerators, then this will most likely mark the end of direct dark matter searches. This unfortunate outcome would inevitably put the last nail in the coffin of the broad class of dark matter candidates that go under the name of WIMPs,[64] and to the hope that the dark matter problem can be cured with the same therapy as that devised to address the pathologies of the Standard Model of particle physics.

But the news of the death of the WIMP patient would be met with joy by the supporters of other dark matter candidates. A sizable fraction of the scientific community is convinced that too much focus has been put on WIMPs at the expense of other perfectly viable candidates. Among these, one that is particularly appreciated by theoretical physicists is the *axion*. Michael Turner, the inventor of the acronym "WIMP", famously said that

> Axions are the thinking persons' dark-matter candidate.

Interestingly, the existence of axions and several other popular candidates, including sterile neutrinos, as we shall see in the next chapter, will be tested by means of dedicated experiments and astrophysical observations on a timescale similar to that needed to build and operate the direct detection tonne-scale experiments.

One of the most interesting experiments currently searching for dark matter axions is the ADMX experiment, at the University of Washington, which aims to detect the light produced by axions when they are in the presence of a strong magnetic field. This experiment is expected to probe the most interesting range of axion masses within the next five to ten years.

Going back to WIMPs, a more optimistic outcome would be that direct detection experiments *do* find evidence for dark matter. At that point, it would be crucial to combine the information

---

[64] See the definition of WIMPS on page 10, under "Warding off the unknown" in Chapter 1.

arising from these experiments with that obtained at the LHC. It is in fact possible that in the "nightmare scenario", where no signs of new physics are discovered at the LHC, dark matter might instead produce a signal in a direct detection experiment. This is possible even in the simplest extensions of the Standard Model! In that case, tonne-scale experiments would represent a unique window for discovering particles that would otherwise go undetected in accelerator searches.

The most exciting possibility is, of course, that dark matter particles are found both in direct detection experiments *and* at the LHC. In this "dream scenario", a lot could be learnt about the nature of dark matter. Both types of experiments, for instance, allow a determination of the mass of the particle, so one could perform an initial check to see whether the two measurements were compatible, and therefore whether the particles discovered in the two experiments could potentially be the same. Then, one could try to construct particle physics models that explain the accelerator data, calculate the strength of interaction for all of those models, and compare this strength with that measured from direct detection experiments, thereby providing an independent test of the nature of the dark matter particle.

This second test is complicated by the fact that in order to estimate the strength of the interaction, an assumption must be made about the number of newly discovered particles hitting the Earth. Back to our earlier analogy, we need to specify what fraction of the "dark raindrops" falling from the sky is made of the particles discovered at the LHC and what fraction is made of something else, say axions or sterile neutrinos.

Fortunately, very effective strategies have been devised that allow one to perform a combined analysis of the data without making unjustified, possibly wrong, assumptions about the fraction of dark matter in the form of new particles. These strategies reduce, basically, to the assumption that the fraction of dark matter particles in the form of neutralinos is the same locally (in the rain of dark particles falling on the Earth) as it is on average in the universe. This procedure is very powerful because it allows one to validate both the particle physics and the cosmological framework adopted in the calculation; it would provide an accurate measurement of the mass, strength of interaction, and abundance of the dark matter particles.

To summarize the possible outcomes: if direct detection experiments discover dark matter particles, they will dramatically improve the prospects for identifying the nature of dark matter particles. If they fail, and if nothing is found at the LHC or in other experiments searching for alternative candidates, particle physics and cosmology will undergo a dramatic paradigm shift that will require brand new ideas and a complete rethinking of our understanding of the universe.

Fortunately, one last detection strategy remains to be discussed, a strategy that may shed light on the nature of dark matter without the need for any dedicated experiments. It goes under the name of *indirect detection*, and is the topic of the next chapter.

7

———

Signals from the sky

"Data! Data! Data!" he cried impatiently. "I can't make bricks without clay."
Sir Arthur Conan Doyle (1859–1930), *The Adventures of Sherlock Holmes*

On June 16, 1977, a semitrailer traveling eastbound on Interstate 80 near Kewanee, Illinois, crossed over into the opposite lane and hit the front left side of a westbound car, killing the driver and causing minor injuries to the other passengers. The driver was Benjamin W. Lee, head of the theoretical physics department at Fermilab and a professor at the University of Chicago, who was driving to Aspen, Colorado, with his wife and their two children to participate in a scientific meeting.

Steven Weinberg, who would be awarded the Nobel Prize in Physics two years later, and the Fermilab physicist Chris Quigg wrote in an essay shortly after Lee's death

> At the time of his death, Lee was in the midst of a period of enormous creativity. . . . He was just beginning a program of research on cosmology and was delighted with this opportunity to move into yet another field.

Together with Weinberg, Lee had started exploring the astrophysical consequences of new hypothetical particles called *heavy leptons*, whose existence they had postulated about one year earlier, and they were pioneers in the field of astroparticle physics, which consists in the application of particle physics methods to astrophysical and cosmological systems. Several other brilliant scientists joined Lee in this study, although Weinberg in the end did not sign the final paper from the study: "commitments to axions and related things unfortunately took their toll on his time", as the authors wrote in the acknowledgments.

Among the many interesting ideas discussed in the paper, one had truly far-reaching implications. As we have seen on page 78, under "Prologue" in Chapter 4, dark matter particles can annihilate into particles of the Standard Model, but when the annihilation rate drops below the expansion rate of the universe, the reactions that turn dark matter particles into Standard Model particles are not effective anymore, and the number of dark matter particles then remains constant.

Lee and collaborators argued that although on average the rate of annihilations in the universe is very small, some *residual annihilations* could take place in overdense regions of the universe such as galactic halos, including our own. They went on to suggest that the light produced by the annihilation of dark matter could be observed with gamma-ray telescopes.

Their preliminary estimates were confirmed by the astrophysicist Floyd Stecker and then applied to many other types of dark matter, including supersymmetric ones. The field of *indirect dark matter searches*, which aims to discover dark matter by searching for the products of its annihilation or decay, was born.

It is impossible to do justice here to the many brilliant scientists who sparked the particle dark matter revolution in the first half of the 1980s: in five years, the theoretical foundations of direct and indirect dark matter searches were laid, and the race for discovery, almost four decades later, is still on.

The pale light of dark matter

In 2009, I gave a public lecture at the Institute of Astrophysics in Paris, titled "La pâle lueur de la matière noire", French for "The pale light of dark matter", where I described the status of indirect dark matter searches. A young boy, about eight years old, approached me after the presentation and asked whether it was still appropriate to call it dark matter if it really emits some form of light. In French, the question makes even more sense, since *matière noire* literally means "*black* matter".

He was right; indirect detection may actually allow us to *see* dark matter. Not with normal telescopes, though: the light produced by the annihilation of dark matter consists in fact of high-energy photons, called gamma rays, whose energy is comparable to the mass of the dark matter particles. In the case of WIMPs, this mass is between approximately ten and ten thousand times the mass of a proton, or between 10 GeV and 10 TeV.[65]

At these energies, photons cannot penetrate the Earth's atmosphere, as they interact with the nuclei of the atmosphere and generate a cascade of other particles. Gamma-ray detectors must therefore be placed above the atmosphere on satellites orbiting around the Earth; alternatively, they must be able to reconstruct the energy and direction of the photons from the properties of the shower of particles produced by their interaction with the atmosphere.

Among the gamma-ray telescopes currently in operation, the NASA Fermi satellite is probably the most suitable for searching for dark matter annihilation. This satellite was launched in 2008 and was placed in orbit around the Earth at an altitude of 550 kilometers. It is 2.5 meters high by 2.8 meters deep and wide, and it detects gamma rays by converting them into pairs of electrons and positrons, using a technique similar to that adopted by the detectors at the LHC.

Ground-based telescopes make use of a completely different strategy to detect gamma rays. The shower of particles produced by the interaction of a photon with the nuclei of the atmosphere produces a characteristic glow, which originates from the fact that the secondary particles propagate in the atmosphere faster

---

[65] For a discussion of the equivalence between mass and energy, see page 88, under "The ring" in Chapter 5.

than light does.[66] This leads to the emission of radiation known as Cherenkov light, from the name of the Russian scientist who measured it for the first time in 1934.

There are several "Cherenkov telescopes" currently in operation, including the HESS telescope in Namibia, the MAGIC telescope in the Canary Islands, and the VERITAS telescope in Arizona, and a large number of scientists worldwide are working together on the construction of a much larger telescope, called CTA, that will consist of an array of hundreds of individual telescopes of various sizes. All of these experiments are expected to detect, or at least to strongly constrain, the flux of photons produced by the annihilation of dark matter particles.

The rate at which dark matter particles annihilate, and therefore the rate at which gamma rays are produced, increases strongly with the density of dark matter. More precisely, it increases with the *square* of the density: if the density goes up by a factor of ten, for instance, then the annihilation rate goes up by a factor of $10^2 = 100$. The flux of gamma rays received from a telescope decreases with the square of the distance from the source, however: if the distance from an overdensity of dark matter is increased by a factor of ten, then the measured flux of gamma rays is *reduced* by a factor of $10^2 = 100$.

The detectability of dark matter halos depends, therefore, on a combination of density and distance. Clusters of galaxies, for instance, are known to host very large concentrations of dark matter, so the annihilation rate will be very high, but they are very distant, so the resulting flux of gamma rays turns out to be relatively small. The dark matter substructures in the Milky Way are much closer, but the flux of gamma rays is still very small in most cases because they host a comparatively small amount of dark matter.

---

[66] As we have seen in Chapter 2, light propagates at different speeds in different media. It is the speed of light in vacuum that provides an upper limit to the speed of any moving entity.

The most promising targets for indirect detection of dark matter—in other words, the direction in the sky in which we need to point our telescopes in order to maximize the chances of discovery—can be identified by means of a careful analysis of the distribution of matter in numerical simulations of the Milky Way.[67] In 2008, Enzo Branchini, Lidia Pieri, and I decided to visualize how the sky would look if we could "see" the faint gamma-ray emission produced by the annihilation of dark matter. Similar studies had already been performed, but we were interested in visualizing the effect of all substructures in the Galaxy, including those which were too small to be observed in numerical simulations. With a combination of analytical and numerical calculations, we obtained the full-sky map shown in Figure 7.1, in which the center of the Galaxy is at the center of the map.

There are, however, many standard astrophysical sources of gamma rays, such as pulsars, active galactic nuclei, and a diffuse

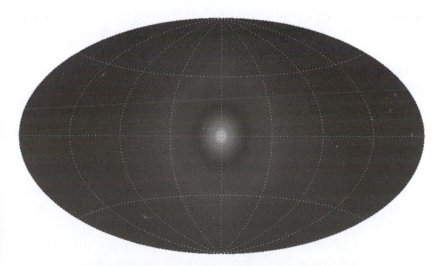

Figure 7.1: A simulated full-sky map of the gamma-ray emission from the annihilation of dark matter particles in the Galaxy. The center of the Galaxy is at the center of the map.

[67] See page 84, under "The plot" in Chapter 4.

Figure 7.2: Full-sky map constructed from two years of observations by NASA's Fermi telescope.

emission due to the gas in our Galaxy, that outshine the faint glow of dark matter particles, as is evident from the fact that a *real* map of the sky in gamma rays, such as the one obtained from the Fermi satellite shown in Figure 7.2, clearly does not resemble the annihilation map in Figure 7.1 at all.

Now, the question is: is there any direction in the sky in which the dark matter signal might exceed that of standard astrophysical sources? A careful comparison reveals that there are indeed "target regions" that may allow effective discrimination of the two signals:

- *The Galactic center*. Located in the constellation Sagittarius, the central region of our own Galaxy constitutes an optimal combination of density and distance. It can be easily identified as the most prominent feature at the center of Figure 7.1. Unfortunately, not only does the same region also host a strong concentration of astrophysical sources (see Figure 7.2), but the actual strength of the annihilation signal also depends strongly on the assumptions that one makes about the distribution of dark matter on scales smaller than those resolved in numerical simulations, a circumstance that complicates significantly the identification of a dark matter signal.

- *Galactic substructures.* Numerical simulations predict the existence of many substructures inside the dark matter halo of the Milky Way (see page 84, under "The plot" in Chapter 4). These appear as small spots in the annihilation map in Figure 7.1. Although it is impossible to know exactly where these substructures are in the sky,[68] reliable predictions can be made for all other properties of this class of objects, such as their number and luminosity. The brightest objects are expected to be associated with the so-called dwarf galaxies, faint systems containing a small number of stars which seems to be almost completely dominated by dark matter. It is by searching for an excess gamma-ray emission from dwarf galaxies that we can set the most robust limits on the properties of dark matter particles.

- *The Galactic halo.* The enhancement of the annihilation signal towards the innermost regions of the Galaxy is rather broad compared with the astrophysical signal shown by the Fermi map. It is therefore possible to search for gamma rays from dark matter annihilation by looking slightly off, say by a few degrees, from the center of the Galaxy. This provides more robust but less stringent limits on the annihilation rate of dark matter particles.

Although gamma rays from dark matter have not yet been identified, there are some intriguing hints. The most recent one is a gamma-ray line, a curious accumulation of photons collected by the Fermi satellite from the direction of the Galactic center at an energy of 130 GeV. It is as if there were particles with a mass equal to this energy that annihilate directly to photons, a feature that was predicted a long time ago as a "smoking gun" for particle dark matter. While the Fermi collaboration is performing an accurate analysis of this claim, other ground-based experiments such as the recently inaugurated HESS-II and the upcoming CTA can be expected to clarify whether the line actually exists or whether, for instance, it is due to some instrumental effect.

---

[68] The positions of the substructures in Figure 7.1 are the result of a particular numerical simulation; if we had chosen another simulation, the positions would have been different.

Antimatter messengers

In the summer of 2008, the astroparticle community was set aflame by the announcement that an anomaly had been observed in the energy spectrum of positrons, the antimatter version of electrons, measured by the Italo-Russian satellite PAMELA.

Even before the actual results were published, the PAMELA data were already circulating in the community, thanks to a trick that was vividly described in an article published on September 2, 2008 in the News section of the journal *Nature*:

> An Italian-led research group's closely held data have been outed by paparazzi physicists, who photographed conference slides and then used the data in their own publications.

There was no wrongdoing on the part of the "paparazzi physicists", who are actually well-respected members of the astroparticle community: they had requested permission to use the data and, in their paper, they acknowledged the PAMELA collaboration and the conference where the data had been presented. But the anecdote reveals the degree of excitement, or perhaps hysteria, caused by the PAMELA discovery in the data-starved field of dark matter searches.

What made this experimental result so interesting is that theoretical physicists had predicted that in presence of dark matter annihilation, the relative abundance of positrons with respect to electrons, also known as the *positron ratio*,[69] should increase with energy, which is precisely what the PAMELA collaboration found.

Although the universe we live in is mostly made of normal matter, with only small traces of antimatter, cosmic rays typically exhibit a larger proportion of antimatter, which is generated in a number of astrophysical processes. However, predictions made on the basis of known astrophysical models suggested that the positron ratio should be a decreasing function of energy. Dark

---

[69] More precisely, the positron ratio is defined as the number of electrons divided by the sum of the numbers of electrons and positrons.

matter annihilation events, in contrast, were known to produce "bumps" in the positron ratio, with a slow rise followed by a rather abrupt decline at an energy equal to the mass of the dark matter particle. The PAMELA result seemed to support the latter hypothesis, and therefore to provide evidence for dark matter particles.

Unfortunately, several shortcomings of this interpretation soon became apparent. The annihilation rate required to explain the data was so high that it should have already led to all sorts of other measurable signals, like an intense gamma-ray emission from the Galactic center, an associated flux of antiprotons, and so on. None of these has been found. Furthermore, the astrophysicists started to revise their calculations, and they realized that, well, there were after all models that could lead to a rising positron ratio.

The careful reader might have noticed that the dark matter signature in the positron ratio had been predicted to be a "bump", whereas the PAMELA results show only a rise in the positron ratio up to the maximum energy measured by the instrument, not the predicted abrupt decline. The detection of the latter feature might be possible with AMS-02, an experiment that was transported by NASA's Space Shuttle *Endeavour* to the International Space Station in May 2011 (Figure 7.3).

AMS-02 has already detected tens of billions of particles,[70] and it is expected to extend the positron ratio measurements to much higher energy. We don't know whether it will discover new features, but we can already predict that the debate over whether the position ratio measurements provide conclusive proof of dark matter will not be settled easily.

The difficulty of distinguishing a dark matter signal from one produced by an ordinary astrophysical source is an inherent disadvantage of indirect searches. To overcome this difficulty, it is crucial to focus on "smoking-gun" signatures of dark

[70] The number of cosmic rays measured by AMS-02 can be checked in real time at <http://ams.nasa.gov/>.

Figure 7.3: The AMS-02 experiment, attached to the International Space Station.

matter: observations that would point directly to the existence of new particles.

We mentioned gamma-ray lines in the previous section; let us now take a look at an alternative signature that is currently being searched for in the depths of the Antarctic ice sheet.

Flashes in the ice

A "dark rain" falls on the solar system from the Galactic halo. In the previous chapter, we have seen how direct detection experiments aim to detect some of these particles showering on the Earth, and perhaps to detect a modulation in the rate of collisions of dark matter particles in underground detectors due to the motion of the Earth around the Sun.

This "dark rain" also falls on the Sun. Since the Sun is much larger than the Earth, dark matter particles are much more likely to collide with one of its nuclei. Some of these particles will lose so much energy in a collision that they will remained trapped by the gravitational force of the Sun. These particles are doomed: they will continue orbiting in and around the Sun until they strike another nucleus, losing even more energy. Gradually, they will lose all of their energy and will inexorably be swallowed by the Sun.

This process is quite rare for an individual particle, but there are many particles passing through the Sun, and an actual calculation shows that, for typical values of the strength of the interaction of dark matter particles with nuclei, a reservoir of dark matter particles will slowly build up at the center of the Sun. The density of dark matter particles captured by the Sun is not high enough to alter the properties of the Sun, but it is sufficient to enhance the annihilation rate substantially. Of all the annihilation products, only neutrinos can escape from the center of the Sun, since any photons and antimatter generated in the annihilation process are quickly absorbed by the surrounding material (Figure 7.4).

If we could measure these neutrinos emerging from the core of our star, we could prove that dark matter particles exist, and identify or at least constrain their nature. The problem now is that neutrinos themselves are very hard to detect. Many experiments exist today for the detection of low-energy neutrinos, such as those produced by nuclear reactions in the Sun or by nuclear

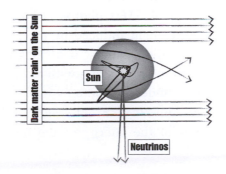

Figure 7.4: High-energy neutrinos might be produced by the annihilation of dark matter particles trapped in the Sun.

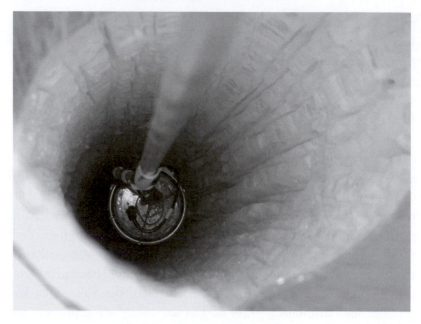

Figure 7.5: Construction of the IceCube neutrino telescope: a photomultiplier is lowered into the depths of the Antarctic ice sheet.

reactors. But no experiment has yet succeeded in detecting *high-energy* neutrinos from astrophysical sources, nor those that might arise from the annihilation of dark matter particles in the Sun.

The IceCube experiment (Figure 7.5) might change all that. Located between 1500 and 2500 meters below the geographic South Pole, it consists of 5160 photomultipliers, devices that can detect the flashes of light induced by the passage of neutrinos through the ice. Although most neutrinos travel unnoticed through the Earth without interacting with any form of matter, a small fraction of them undergo reactions with the ice surrounding the detector that convert them into *muons*.[71] These particles carry a large fraction of the energy of the neutrino, and they

---

[71] Muons are the "heavy cousins" of electrons, and we have already encountered them in Chapter 5; see in particular page 91, under "The ring".

propagate through the ice faster than light does, causing the emission of flashes of Cherenkov light, which can be observed by light detectors.

This detection technique was pioneered by Russian scientists in an experiment in Lake Baikal, an immense freshwater lake in Siberia, which has the virtue of being one of the deepest and cleanest lakes on the Earth. Since then, several other experiments have been built in the Mediterranean, off the coasts of France, Italy, and Greece, and there are plans to build an even larger neutrino telescope, Km3Net, in one of these three locations.

Although cosmic neutrinos have not been detected yet, these experiments have been useful for a variety of biological studies. In particular studies have been performed of some bacteria and other species that produce light in the depths of the ocean, such as aquatic fireflies. The songs of whales can also be recorded by the microphones installed around the telescopes to pick up the faint sound waves produced by neutrino interactions. We will come back to the broader impact of particle physics and astroparticle experiments in the next chapter.

Black holes and other monsters

Radio Alternantes is a nonprofit radio station in Nantes, France. In 2009, on the occasion of a conference on dark matter at the Nantes Museum of Natural History, a journalist asked me whether I would be available for a walking radio interview: the idea was to discuss dark matter and science in general as we walked through the streets of Nantes from the train station to the Museum, and to randomly stop passersby from time to time to ask their opinions and stimulate questions.

I gladly agreed, and so we spent about one hour strolling around Nantes, talking about science with, literally, women and men on the street. Among the people we met, who included a philosopher who correctly guessed the fraction of the energy density of the universe in the form of baryons (!), several asked the question

Dark matter? Does it have something to do with black holes?

The association is inevitable, I guess, for nonexperts, since both terms evoke darkness, and perhaps also mystery, and it turns out that this is indeed an interesting question. There are strong arguments against the possibility that dark matter is made of black holes, but it is possible that the formation of these extreme objects may have an effect on the distribution of dark matter.

Black holes can be broadly divided into three different classes, as illustrated schematically in Figure 7.6. The first includes *stellar-mass* black holes, with a mass smaller than one hundred times the mass of the Sun.

There is robust evidence for these objects, which constitute the endpoint of the life of massive stars. We observe stellar systems consisting of two bodies orbiting around each other; in some cases the mass of one of these objects is larger than the maximum mass that can be achieved by a compact object, that is, the mass above

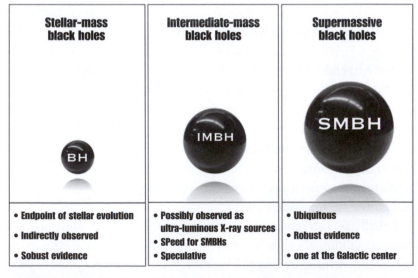

Figure 7.6: Properties of stellar-mass, intermediate-mass, and supermassive black holes.

which a compact star cannot withstand its immense gravitational pull and it therefore implodes to a black hole.

The third class of objects shown in Figure 7.6 is that of *supermassive* black holes, with a mass between a million and tens of billions times the mass of the Sun. These objects are now observed in many galaxies, and it is believed that in fact a supermassive black hole exists at the center of *every* galaxy. The most striking evidence for these objects can be found in our "Galactic backyard", right at the center of our own Galaxy, as we have seen in Chapter 4.

Another class of object could exist between stellar-mass and supermassive black holes, with a mass between a hundred and a million times the mass of the Sun: *intermediate-mass* black holes. Although there is circumstantial evidence for these objects from a variety of astrophysical observations, their existence remains speculative at this stage.

None of these objects can constitute the dark matter in the universe: we would have already detected them through microlensing, i.e., distortion of the light from background stars, and other astrophysical measurements. But black holes can have a dramatic effect on dark matter: they can act as powerful *annihilation boosters*, by increasing the density of dark matter around them.

To understand the physical mechanism underlying this effect, let us consider for a moment the Sun–Earth system. Imagine that at some point we start to increase the mass of the Sun, on a timescale much longer than the Earth's revolution period. What would happen? The Earth would conserve its angular momentum around the Sun, but its orbit would become smaller, in response to the increased gravitational potential.

Now imagine a distribution of dark matter around a black hole that is growing in mass by swallowing astrophysical junk, such as blobs of gas, stars, and so on. As the mass of the black hole increases, the distribution of dark matter shrinks, becoming more compact and more dense.

The formation of black holes therefore enhances the annihilation rate of the surrounding dark matter, but only under the condition that the growth of the black hole is slow. This excludes the possibility of observing any effect around stellar black holes, which form violently from the collapse of stars. Supermassive black holes are more interesting from this point of view, but unfortunately the dense concentrations of stars surrounding them tend to destroy any overdensity of dark matter.

Joe Silk, who cosupervised my Ph.D. with Günter Sigl, first at the University of Oxford and then at the Institute of Astrophysics in Paris, pioneered the field of the effect of black holes on dark matter together with Paolo Gondolo with a study of the effects of the growth of the supermassive black hole at the Galactic center. During my Ph.D. defense, back in 2003, Joe asked only one question: "What about other black holes?"

It took me more than a year to identify a particular type of black hole—known as intermediate-mass black holes—that could strongly modify the distribution of dark matter around them and act in practice as dark matter annihilation boosters. The idea arose from discussions with Andrew Zentner, who was a postdoc at the University of Chicago when I was working at Fermilab, and who by chance happened to live in the same building as me in the South Loop area of Chicago.

The scenario we proposed, together with Joe Silk himself, suggests that it could be possible to observe as many as several tens of these objects as *unassociated gamma-ray* sources, i.e., sources of high-energy photons or neutrinos that, unlike astrophysical sources, do not emit any other type of radiation. This is perhaps a speculative scenario, but one that if confirmed would teach us a lot about the nature of dark matter and black holes.

Besides black holes, dark matter can in principle interact with other stellar objects. If the density of dark matter is high enough, such as at the center of galactic halos or in the primordial minihalos in which the first stars formed, it is possible that the rate of annihilation of dark matter particles may become so high that the energy injected into the cores of the stars located in these extreme environments modifies their structure and evolution.

Many other ideas have been proposed for observing dark matter particles indirectly through their effects on the heat flow of planets, including the Earth, on the luminosity of galaxies and galaxy clusters, and on cosmological observables such as the cosmic microwave background. As the epigraph of this chapter implies, only observational *data* can settle the debate, enacting once more what T. H. Huxley defined as[72]

> The great tragedy of Science—the slaying of a beautiful hypothesis by an ugly fact

or paving the way to a new golden age of astroparticle physics.

---

[72] T. H. Huxley, "Biogenesis and Abiogenesis", in *Collected Essays*, vol. 8 (1893–1894).

8

The road ahead

The moment of truth

We like to think of the history of science as a perfect, rational, ever-increasing progress from the abyss of ignorance to ever-loftier heights of knowledge. It's not. Or, at least, not always.

Looking back at the historical development of scientific knowledge, a clear increasing trend in scientific progress can only be identified on a timescale of decades or centuries. On much longer timescales, science proceeds in cycles where "golden" and "dark" ages alternate, which closely follow the rise and fall of civilizations and empires. If, for instance, the advances in modern science since the time of Galileo Galilei and Isaac Newton, who lived more than two centuries ago, are many and indisputable—think of the countless discoveries ranging from fundamental particles up to the largest cosmological scales—there are plentiful examples of past civilizations whose scientific knowledge faded out or froze abruptly, only to be rediscovered after many centuries, as in the case of the ancient Greek and Islamic civilizations.

On timescales shorter than a few decades, the most relevant for *frontier* science, the dynamics of scientific research is much more complex. Rather than a straight exemplification of the scientific method, it resembles more a Conan Doyle novel: hypotheses are proposed and discarded, observational results turn into misleading clues, and unexpected *coups de théâtre* suddenly change the course of the action.

This is where the elegance and precision of the scientific method sometimes give way to other aspects of human nature. Prejudice, stubbornness, insecurity, and envy, as well as practical considerations related to academic careers and prestige, sometimes take their toll on the short-term efficacy of the scientific method. But at least the long arms of politics and financial interests, which so dramatically affect many other human activities, rarely extend to the field of fundamental research, leaving scientists, with all their strengths and weaknesses, to resolve their disputes among themselves.

Fortunately, the rigorous implementation of the scientific method allows, at least in the long term, even the most fierce debates to be settled. One famous example is the controversy surrounding the existence of the *aether*, a substance that was thought to pervade the universe and to be the "sea" in which electromagnetic waves propagate. In the eyes of many scientists of the 18th and 19th centuries, it was a perfectly reasonable hypothesis—very much like, we might be tempted to say, the ideas of dark matter and dark energy today.

The origins of the term "aether" are shrouded in the fog of time. It was certainly in use well before Aristotle, since in his cosmological treatise *On the Heavens* he wrote:[73]

> The common name [aether] has been handed down from our distant ancestors. . . . The same ideas, one must believe, recur in men's minds not once or twice but again and again. And so, implying that the primary body is something else beyond earth, fire, air, and water, they gave the highest place a name of its own, aether, derived from the fact that it "runs always" for an eternity of time.

Ironically, the concept of the aether indeed recurred again and again in the minds of scientists for two millennia after Aristotle. It was used by medieval alchemists, under the name

[73] Aristotle, *On the Heavens*, translated by J. L. Stocks, <http://ebooks.adelaide.edu.au/a/aristotle/heavens/>.

of *quintessence* or *fifth element*, and it was featured in the philosophy of Kant and Descartes. In one of its later incarnations, it took the name of *luminiferous* aether, i.e., "bearer of light", as it was thought to constitute the medium in which electromagnetic waves propagated.

As we shall see below, by the early 20th century the concept of the aether had been convincingly ruled out: the sophisticated experiments that had been devised to test for its existence and measure its properties had found *nothing*, and new theoretical advances, in particular Einstein's theory of special relativity, had in any case made it an obsolete concept.

Quite the opposite was the case for the Higgs boson, as we have seen. Proposed in the 1960s, this particle, and its associated field, has been on the agenda of at least two generations of physicists. Various experimental searches have, year after year, constrained the possible values of its mass.

The skeptics proposed ways of getting rid of it. The contrarians criticized the search strategies. Two overly anxious men even pursued a lawsuit in a Federal court in Hawaii alleging that accelerator physicists were threatening the safety of our planet. And then, in 2012, the Higgs boson, or at least a particle that closely resembles it, was found by the ATLAS and CMS collaborations at the LHC.

It is instructive to look at the current status of modern physics and astronomy in the light of these historical considerations. As we have seen in this book, physicists and astronomers have come to accept the existence of an unseen form of matter to explain the motion of celestial bodies, and proved convincingly that it must be made of new, as yet undiscovered, particles.

The main ideas behind the extraordinary theoretical and experimental efforts that are currently being made towards the identification of dark matter were laid down in the 1970s and early 1980s, including the possible connection between the problem of the missing mass in the universe and the new particles proposed by particle physicists searching for a fundamental theory of all particles and interactions. Today, the time has come to test this conjecture. By the end of the decade, CERN's particle accelerator, the Large Hadron Collider, will have accumulated several years of data at the maximum energy after a shutdown period in 2013 and 2014, allowing us to probe a very large fraction of the most widely discussed dark matter models. And, on a similar timescale, the next generation of astroparticle experiments is expected to deliver complementary information about the nature of dark matter particles.

In other words, the moment of truth has come for many dark matter candidates, especially those falling into the broad category of weakly interacting massive particles. Thirty years have elapsed since these particles were proposed as the solution to the dark matter problem, and at least two generations of physicists have worked out detailed predictions for a wide array of experimental searches and built ever-larger and more sensitive experiments to test those predictions. We must now either discover them or rule them out.

The nightmare scenario

The unfortunate case where no new particle besides the Higgs boson is discovered at the Large Hadron Collider is sometimes referred to as the "nightmare scenario" of particle physics. That's because the Higgs is the last missing piece of the Standard Model of particle physics, whereas many hope that accelerator data will also provide a glimpse of what lies *beyond* it, whether that is related to dark matter or not. There is no doubt that failing to discover new particles at the Large Hadron Collider would be a big disappointment for many, but fortunately that would not be the end of physics. There would still be ways to discover new particles, even in the form of weakly interacting massive particles.

First, the discovery of the Higgs represents an extraordinary opportunity to perform precision tests of the Standard Model. Any theory that seeks to extend it, for instance supersymmetry, predicts subtle differences in the number of Higgs bosons produced and in the way this particle decays into lighter particles.

A hint of a possible deviation from the predicted rate of decays of the Higgs boson into a pair of photons has actually been observed recently, possibly signaling the existence of new particles involved in the decay process. Precision studies at the LHC and possibly at future accelerators will tell us whether the Higgs boson is actually that predicted by the Standard Model or whether its properties are better explained by some new theory.

Another reason to remain optimistic, even in the absence of new discoveries at the LHC, is that it is possible that astroparticle experiments will discover dark matter particles. As we have seen, underground *direct detection* experiments aim to reach a detector mass of about one tonne in the next few years, increasing the sensitivity to interactions between dark matter and ordinary matter by a factor of ten or more.

Even in the simplest extensions of the Standard Model, such as *constrained* supersymmetric models, in which many assumptions are made to simplify the structure and improve the predictive power of an otherwise extremely complex theory, it is possible that dark matter particles that escape detection at the LHC will instead be found in direct detection experiments. The identification of dark matter might be difficult at that point, in the absence of any collider signature, but a combination of different direct detection experiments might provide complementary information, as we shall discuss in the next section. There is no doubt that a *tailored* accelerator would then become a high priority of the international community.

Astrophysical observations might also eventually come to the rescue. As we have seen, an excess of gamma rays with an energy of 130 GeV has been observed from the direction of the Galactic center with NASA's Fermi satellite. If confirmed, this would be a signal crying out for an explanation that is hard to find with conventional astrophysical models. Similarly, if high-energy neutrinos from the Sun are observed, this might also point directly to a dark matter explanation, as discussed in the previous chapter.

What if not even astroparticle experiments find evidence for weakly interacting particles? There are other options on the table, but it is unclear to what extent they are really viable, or how long they will remain viable. Axions and sterile neutrinos are the best alternatives to weakly interacting massive particles as dark matter candidates, but in both cases experiments are closing in on these particles, exploring the most favored range of masses and interaction strengths.

As a last resort, many physicists would probably try to explore new dark matter candidates, but many others would probably try to get rid of dark matter altogether, going back to theories based on *modified gravity*. Instead of assuming Einstein's theory of general relativity and adding new unseen matter to reconcile theoretical predictions with observations, they may try to argue that there is only visible matter in the universe and that the discrepancy with the observations arises from the failure of Einstein's equations to explain the behavior of gravity at large distances.

The most important thing, at this point in the history of physics and astronomy, is that we are about to get some answers. Even in the worst-case scenario discussed in this section, the impact of these new results on our understanding of the universe would still be profound and worth the effort. Realizing that one is on the wrong track in science, just as in any other human adventure that carries risks, is as important as success.

An extraordinary example of the importance of null searches is provided by a famous experiment performed by Michelson and Morley in 1887. As we have seen above, most scientists were convinced back then that electromagnetic waves propagated in a medium known as the aether. But when the American physicist Albert A. Michelson set out to measure the effect of the relative motion of the Earth with respect to the aether on the speed of light, he failed to find any effect.

Michelson then asked for the assistance of Edward W. Morley to perform a better, more precise experiment, and they jointly built a device—the Michelson–Morley interferometer—which successfully reduced the largest sources of experimental error that Michelson had to face, such as vibrations and various other sources of distortion. This was achieved thanks to a pool of mercury, on top of which the experiment could float and rotate uniformly.

The aether theory predicted that as the experiment changed its direction with respect to the supposed aether (it rotated once every six minutes around its center), the position of the pattern of light produced by a lamp—more precisely, the position of the fringes produced by the interference of light beams along two different optical paths that included portions orthogonal to each other—would exhibit a characteristic "sinusoidal" behavior. That is, the pattern would move first to one side, return to the zero position, and then move to the other side by an equal amount.

The predicted result is shown as a dashed line in Figure 8.1, which is taken from the actual article published by Michelson and Morley in the *American Journal of Science* in 1887. Much to their surprise, they measured a much smaller displacement, shown in the figure as a solid line. They repeated the experiment at noon (top panel) and in the evening (bottom), rotating the apparatus in opposite directions, but there was no trace of the predicted signal. In fact, in order to make the observed result visible in the figure, the predicted signal had to be rescaled by a factor of eight! Clearly,

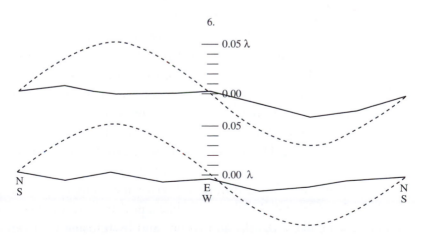

Figure 8.1: Result of the Michelson–Morley experiment.

it was impossible to reconcile the experimental result with the theoretical prediction.

Several brilliant scientists, including Lorentz, FitzGerald, Poincaré, and many others, attempted to find a theoretical explanation for this completely unexpected result—leading, incidentally, to interesting developments in physics—but the aether was doomed. In 1905, the then 26-year-old Albert Einstein published an article titled "On the electrodynamics of moving bodies", in which he laid the foundations of special relativity, and in unassuming terms he declared the aether dead:

> The unsuccessful attempts to discover any motion of the earth relatively to the "light medium", suggest that the phenomena of electrodynamics as well as of mechanics possess no properties corresponding to the idea of absolute rest. . . . The introduction of a "luminiferous ether" will prove to be superfluous.

In a similar way, dark matter particles cannot be *ruled out* by current experiments, only *strongly disfavored*, until perhaps a new theory is proposed that can explain all observations without invoking them, thus making dark matter as superfluous as the aether.

The consequences of success

One of the most famous ukiyo-e—a Japanese word for woodcuts that literally means "images of the floating world"—created by the Japanese artist Hanabusa Itcho (1652–1724) illustrates the story of the blind men and the elephant (Figure 8.2). Many versions of this story exist, but the best-known is perhaps that contained in the Buddhist scripture known as the *Udana*, in which the Buddha tells a parable of six blind men who are asked to describe an elephant.

Each of them touches a different part of the animal: the one who feels the elephant's head says it's like a pot, the one who feels the tail says it's like a brush, and so on, and by refusing to listen to the others and insisting on their own interpretation, they fail

Figure 8.2: Blind Monks Examining an Elephant. Ukiyo-e by the Japanese artist Hanabusa Itcho (1652–1724).

to see the truth. The Buddha explains the meaning of the parable with the following verses:

> O how they cling and wrangle, some who claim
> For preacher and monk the honored name!
> For, quarreling, each to his view they cling.
> Such folk see only one side of a thing.

Suppose now that dark matter is actually discovered. And not just by one but by several, possibly many, experiments. Can we identify, starting from the partial information accessible to each of these experiments, the true nature of dark matter? Will we be able to piece together the puzzle of the particle nature of dark matter from measurements like a rate of events in a direct detection experiment or missing energy in a particle detector?

When I visited Carlos Muñoz and David Cerdeño at the Department of Physics of the Universidad Autonoma de Madrid in March 2008, these questions were at the heart of our discussions.

Cerdeño and I had been discussing an article in which the authors—Michael Peskin, Marco Battaglia, and collaborators—had shown that even if new particles were discovered at the LHC, it would be hard to convincingly identify these particles with dark matter on the basis of accelerator data only.

We thus wondered whether adding information from direct or indirect searches would allow us to obtain a clear, definitive identification of dark matter particles. It took us a couple of weeks to formalize the problem and to identify the technical challenges. The idea was to start from a theoretical model for the nature of dark matter, and then to study to what accuracy the ATLAS and CMS detectors at CERN could measure its properties, as Peskin, Battaglia, and collaborators had done, extend the analysis to direct detection experiments, and reconstruct the properties of dark matter particles from simulated data.

Our goal was to demonstrate that a combination of accelerator and direct detection data would be sufficient to reconstruct with good precision the properties of dark matter particles. There were some difficulties, though: a conceptual and a technical one. The conceptual challenge consisted in the fact that accelerator data indeed directly constrained the properties of new particles, but did not say much about their possible identification with dark matter, whereas direct detection experiments only constrained a combination of particle physics properties and (poorly known) astrophysical quantities. As we shall see below, this obstacle was overcome thanks to the introduction of an assumption that was later called the *scaling ansatz*.

The technical difficulty consisted in the need to perform a statistical analysis that would allow us to take account of all theoretical uncertainties, as well as all experimental errors, and to obtain at the end of the process a tool that would allow us to compare quantitatively the properties of dark matter obtained from simulated data with those of the theoretical model we had started with. We were in fact interested in understanding whether the data we had simulated would have allowed a precise identification of the dark matter particle.

We tackled this technical challenge by first studying the scientific literature, and started writing from scratch a code to solve this problem. But we then realized that an easy solution was to hand, since one of the experts in this type of statistical analysis, Roberto Ruiz de Austri, was a postdoc in the same institute in Madrid, and he was actually sitting in an office just a few doors down the hall. He and Roberto Trotta, now a faculty member at Imperial College London, had written a publicly available code that solved precisely the type of problem we were interested in.

Joining forces with Roberto Ruiz and Roberto Trotta, we formed a collaboration team, which, over the years, saw the participation of several students and postdocs, including Mattia Fornasa, Miguel Pato, Lidia Pieri, and Charlotte Strege. It allowed us to go much further than any of us could have done alone in studying the prospects for identifying dark matter by means of a combination of experimental searches. It is instructive to take a brief look at the possible outcomes of the experimental searches for dark matter if one or more of these searches finds evidence for it. There are several possible scenarios, as discussed below.

NEW PARTICLES ARE FOUND AT THE LARGE HADRON COLLIDER ONLY

The discovery of new particles would be of paramount importance. It's not just the *existence* of new particles that matters— although this would be an extraordinary discovery per se—but also the fact that such a discovery might allow us to achieve a unified description of all interactions and to access a more fundamental understanding of nature. Obviously, these new particles would represent obvious candidates for dark matter, but how do we make sure that's the case?

The first condition that a new particle must fulfill in order to be considered a viable dark matter candidate is that its abundance must match the cosmologically observed value (see the ten-point test on page 58, under "Monsters" in Chapter 3). This means in particular that the particle must be *stable*, or at least that it should not decay into any other particle on a timescale smaller than the age of the universe. Otherwise, any such particles produced in the early universe would by now have decayed away, and could not explain the dark matter observed today.

However, all we can learn from accelerator data is that the particle must be stable over the time it takes to fly from the interaction point where it was produced in a collision of two proton beams to the outer edge of the detectors. For a particle traveling close to the speed of light, crossing an experiment even as big as ATLAS is a matter of a fraction of a second!

In order to extract further information from accelerator data, we need to make some assumptions about the nature of the observed particles—we may, for instance, decide to interpret the data in the framework of supersymmetry and assume that the dark matter particle is the neutralino. Starting from a measurement of particle masses and other experimental quantities, we can perform a set of complex but ultimately straightforward calculations to estimate what fraction of dark matter is in the form of neutralinos. In an ideal world, this procedure would tell us with certainty that 100% of dark matter was in the form of neutralinos, and allow us to validate our particle physics model for dark matter. We could in fact *identify* the neutralino with the dark matter particle!

Unfortunately, this is unlikely to happen. Figure 8.3 shows a more likely outcome of a dark matter identification procedure based on LHC data only: we can at most hope to estimate the *probability* that neutralinos constitute a given fraction of dark matter. However, instead of a sharp probability peaking around a fraction of 100%, we would most probably find a complex probability distribution, with multiple, very broad peaks.

In this example, based on an accurate simulation of a realistic outcome for LHC data, the probability has two peaks, one around 100% and one around 1%. This means that we would not be able to tell whether neutralinos constitute all of the dark matter in the universe or whether they are a subdominant species contributing

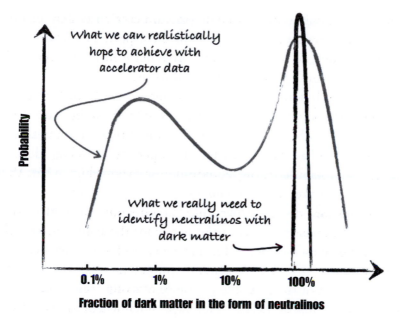

Figure 8.3: Estimate of the fraction of dark matter in the form of neutralinos, using accelerator data only.

only 1% of it, while the rest is in the form of some other particles.

In short, if new particles are found at the LHC, we will learn a lot about their particle physics nature, but it will be difficult to prove beyond reasonable doubt that they constitute all of the dark matter in the universe. Fortunately, complementary information can be obtained from other experiments.

DARK MATTER PARTICLES ARE FOUND IN DIRECT DETECTION EXPERIMENTS ONLY

The case where dark matter particles are found only in direct detection experiments is radically different. Suppose, for instance, that an experiment like Xenon1T detects a number of "hits" substantially larger than the expected number of background events.

In principle, this would be sufficient to claim discovery of dark matter, and even to estimate the properties of the dark matter particles, such as their mass and probability of interaction with ordinary matter.

In practice, the interpretation of the result would be much more complicated. How do we make sure that the measured hits were produced by interaction of dark matter particles with our detector? We have supposed that the number of hits is substantially larger than the expected number of background events, but how do we make sure that our estimates of the background are accurate? How do we prove that the particles we find constitute all of the dark matter in the universe when we can estimate only the product of their local density and the probability of interaction?

In the eyes of many active researchers in the field, a convincing claim of discovery can be made only when two or more experiments find results which are mutually compatible within a well-understood theoretical framework, not necessarily the one most in vogue today, but nevertheless one that can be used to interpret all existing experimental results. It is also necessary that there is no convincing experimental result that contradicts those findings.

In the best-case scenario, where a coherent interpretation of several direct detection experiments can be given and where those experiments use different detection techniques, the following information about the newly discovered particles can be obtained: an estimate of the mass with an accuracy of about 10% if the particles are light, say below 50 GeV, but with a much worse accuracy if the particles are heavier than that, and an estimate of their local density times the probability of interaction, which is again more precise for light than for heavy particles.

In short, a discovery in direct detection experiments *only* would prove the existence of new particles, and allow rough estimates of their mass and their probability of interaction with ordinary matter. But it would hardly allow us to identify the particle physics nature of dark matter particles.

DARK MATTER PARTICLES ARE FOUND IN INDIRECT SEARCHES ONLY

If dark matter particles are found *indirectly*—that is, by measuring the light or secondary particles produced by their annihilation or decay—the prospects for identifying them depend on the specific observational results. It is unlikely that in the absence of complementary information from other searches, featureless "bumps" above the astrophysical background could be convincingly interpreted in terms of dark matter annihilation or decay.

There are, as a matter of fact, many examples in the scientific literature of signals that may or may not be caused by dark matter particles, including observational results from the Fermi, Integral, and PAMELA satellites. For each of these observations, however, there is a possible alternative interpretation in terms of standard astrophysical sources. So how do we convince ourselves that what we are seeing has anything to do with dark matter?

The answer is that we can do so by looking for "smoking-gun" signatures of dark matter, such as the 130 GeV emission discussed in the previous chapter or the observation of high-energy neutrinos from the center of the Sun. In both cases, useful information about the nature of these new particles could be extracted, but a convincing identification with dark matter would still be hard to achieve in the absence of additional information from accelerator data.

In short, just as in the case of direct searches, indirect detection experiments could at best prove the existence of new particles and allow rough estimates of their mass and their probability of interaction with ordinary matter. But they would hardly allow us to identify the particle physics nature of dark matter particles.

COMBINING ACCELERATOR AND DIRECT DETECTION DATA

A much more precise identification of dark matter particles could be achieved by combining the results of two or, even better, all

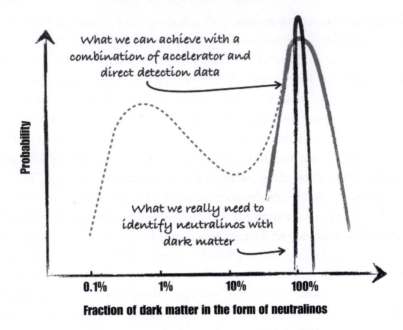

Figure 8.4: Estimate of the fraction of dark matter in the form of neutralinos, using a combination of accelerator and direct-detection data.

three detection strategies. A demonstration of the power of complementarity is shown in Figure 8.4, which shows a reconstruction of dark matter properties from a combination of the accelerator data used in Figure 8.3 and simulated direct detection data.

The probability is now peaked around a dark matter fraction of 100%, and the spurious solution on the left has been robustly ruled out. This is possible thanks to a small trick that my collaborators and I have called the scaling ansatz: when analyzing the combined data sets, we made the assumption that the fraction of dark matter in the form of neutralinos is the same locally—that is, in the neighborhood of the solar system—as on average in the universe. When we came up with this idea, in a particularly enlightening brainstorming session in Madrid, we immediately realized the far-reaching consequences of this seemingly trivial trick: it allows us to rule out the possibility that neutralinos are a subdominant component of dark matter but, thanks to a very large interaction probability with ordinary matter, that they still produce a high number of hits in direct detection experiments.

Making an additional assumption is something that scientists do reluctantly, as it reduces the credibility of the result—the skeptics can always ask "what if the assumption is wrong?" But, fortunately, the price is not very high: my collaborators at the University of Zurich[74] and I have in fact verified that in the case where the remaining component of dark matter (if any) is as cold as the neutralinos, the scaling ansatz is automatically fulfilled.

In short, a self-consistent combination of accelerator and direct detection data may provide enough information to precisely identify the nature of dark matter particles.

So what?

In conclusion, the possible outcomes of the search for dark matter particles are a "nightmare scenario", in which all current experiments yield a null result, and a scenario in which one or more experiments discover new particles, in which case a reconstruction of the properties of dark matter becomes possible.

In the former case, the best we can hope for is that the ideas that led us to hypothesize the existence of new particles and to the experiments constructed to search for them will turn out to be useful for the future of physics, just as the ideas behind the aether and the Michelson–Morley experiment stimulated the ideas of Lorentz, Poincaré, and others, paving the way for Einstein's theory of special relativity. It may seem a small reward for the enormous ongoing effort, but as we have seen at the beginning of this chapter, cutting-edge scientific research does not always follow a linear path. Detours and cul-de-sacs are unavoidable for those who venture to the frontiers of knowledge.

---

[74] Donnino Anderhalden, Juerg Diemand, Andrea Maccio, and Aurel Schneider.

The case where dark matter is eventually discovered, as happened with the search for the Higgs boson, which has been discovered after decades of painstaking searches, is certainly a more desirable outcome of this scientific adventure. It is impossible to overestimate the impact that such a discovery would have on physics and, more generally, on our understanding of our role in the universe. Certainly, particle physics would be revolutionized and understanding the "dark sector" of physics would become a top priority for generations to come.

As we approach the end of our journey, the more pragmatic readers might wonder, "So what?" What might be the impact of either scenario on our everyday life? Why are we investing public money in the quest for this elusive form of matter? These are legitimate questions, which are rarely asked at public conferences—the audience is usually made up of people for whom the subject is interesting per se—but they are sometimes brought up in conversations with people who are not familiar with scientific research or who simply want to understand better the importance of fundamental research.

There are several ways to tackle this question. We might recall the technological applications of accelerator physics, which include medical applications such as PET scans and hadron therapy for the treatment of tumors, and applications in computer science such as globally distributed computing, which finds application in many fields ranging from genetic mapping to economic modeling, and the World Wide Web, which was developed at CERN as a network to connect universities and research laboratories.

We might also stress the importance of these scientific endeavors in the formation of generations of young scientists: thousands of doctoral students undertake a research program in astroparticle physics worldwide every year, learning tools and acquiring skills that they then export to many other disciplines and businesses.

Or we could perhaps even argue that the best reason to engage in fundamental research is that *it's a noble thing to do*, and quote the epigraph of Chapter 3, in which Ulysses incites his companions to venture into unknown land, to advance knowledge and discover new territories.

Robert R. Wilson, the founding director of Fermilab, whose vision and breadth of research interests made him a truly inspirational figure, famously compared scientific research to other forms of human expression such as art and literature on the occasion of a hearing in front of the US Congress's Joint Committee on Atomic Energy. Senator John Pastore—perhaps best known to the public for his role in a hearing that granted 20 million dollars for nonprofit American public broadcasting television at a time when President Nixon wanted to cut expenditure because of the war in Vietnam—asked him to explain why they should spend 250 million dollars to construct a particle accelerator: Would it somehow project the US into a position of being competitive with the Russians? Was it relevant to the security of the country? To which Wilson replied:

> [the construction of the accelerator] has to do with: Are we good painters, good sculptors, great poets? I mean all the things that we really venerate and honor in our country and are patriotic about. . . . In that sense, this new knowledge has all to do with honor and country but it has nothing to do directly with defending our country except to help make it worth defending.

But though the societal and economic benefits of fundamental scientific research are many and important, they are not the main driver for scientists. The truth is that to explore the frontiers of science is an extraordinarily exciting adventure. The possibility of discovering new particles and new facts about the fundamental nature of the universe is a source of thrills and excitement that greatly compensates for the difficulties inherent in an academic career.

The quest for knowledge resonates with something deep inside us, and the feeling of pondering the secrets of nature is exquisitely invigorating, as perhaps best described by an epigram attributed to the astronomer Ptolemy, who lived in the 2nd century AD but apparently experienced the same feelings as modern scientists do:[75]

> I know that I am mortal by nature and ephemeral, but when I trace at my pleasure the windings to and fro of the heavenly bodies, I no longer touch earth with my feet. I stand in the presence of Zeus himself and take my fill of ambrosia.

[75] O. Gingerich, *The Eye of Heaven: Ptolemy, Copernicus, Kepler*, Springer (1997), p. 4.

List of figures

Index

Index